ARCHITECTURAL INTERIOR SYSTEMS

Lighting, Acoustics, Air Conditioning

Third Edition

John E. Flynn
Jack A. Kremers
Arthur W. Segil
Gary R. Steffy

VNR **VAN NOSTRAND REINHOLD**
_____ **New York**

Library of Congress Catalog Card Number 91-3769
ISBN: 0-442-00264-5

Printed in the United States of America.

Van Nostrand Reinhold
115 Fifth Avenue
New York, New York 10003

Chapman and Hall
2-6 Boundary Row
London SE1 8HN, England

Thomas Nelson Australia
102 Dodds Street
South Melbourne 3205
Victoria, Australia

Nelson Canada
1120 Birchmount Road
Scarborough, Ontario MIK 5G4, Canada

16   15   14   13   12   11   10   9   8   7   6   5   4   3   2   1

Library of Congress Cataloging-in-Publication Data
Architectural interior systems : lighting, acoustics, air conditioning
      / John E. Flynn . . . [et al.].—3rd ed.
         p.   cm.
      Includes bibliographical references and index.
      ISBN 0-442-00264-5
      1. Buildings—Environmental engineering.   I. Flynn, John E. (John
Edward), 1930-1980.
      TH6021.A68   1991
      729'.2—dc20   91-3769
                                                             CIP

Contents

Preface

Like the first and second editions, this book is concerned with aspects of building-system engineering and design that affect human sensory response, behavior, productivity, and impressions of well-being. It recognizes the fact that perception and appreciation of a space by an individual or group is dependent, in part, on the interaction of various forms of environmental energy—light, sound, heat—that are precisely controllable by mechanical means.

Unlike the second edition, this third edition has entirely rewritten text and revised graphics in the thermal chapters (chs 3 and 7). Professor Jack Kremers of Kent State is responsible for this improvement, and brings this text up to date for architecture, interior design, and engineering students of the 1990s. Only the most current, energy-effective architectural interior systems are reviewed.

The importance of environmental systems to human experiences, particularly in terms of sensory impressions such as comfort, well-being, and appropriateness, as well as behavioral experiences described by terms such as attention, path selection, assembly, recreation, relaxation, meditation, work, and other forms of human participation has been increasingly accepted in the last decade. Whether this constitutes an applied-psychology problem, an architectural problem, or a human-factors industrial-engineering problem seems a matter of semantics. Most certainly, consideration of human sensory response and behavior has fallen within the scope of architectural design, and thus the subject and its implementation techniques have relevance for anyone concerned with the process, method, and purpose of architectural and interior design.

But knowledge of this general subject is complicated by the rapid rate of technological and scientific development that quickly makes training, design data, and scientific techniques obsolete. This characteristic has become more pronounced in recent years as the compounding effects of limited energy for buildings, increasing building-equipment costs, and installation and maintenance labor costs have combined to introduce significant changes in both criteria for system performance and in the selection of equipment and devices.

Rather than providing a collection of comprehensive engineering details and data, then, the objective of this book is to develop an *overview* of the subject—a sense of architectural perspective, as well as positive guidelines for professional judgment. For this reason, when calculations are shown or implied, they are generally limited in scope and depth to those aspects and ideas that are useful for *preliminary* system analysis. Some background information on evolving building energy limits is also provided.

This book is designed to serve as a general reference for architects, interior designers, and consulting specialists who work in multidisciplinary design-management areas. It is also intended to serve as an introductory text for architectural-design, interior-design, and architectural-engineering students. It offers a concentrated interpretive study of the purposes and functions of buildings in terms of currently accepted sensory needs of occupants, building technology, and energy-demand limits. In scope, it is intended to *bridge* the technology of environmental control and the art of design—to recognize environmental energy as a potential creative medium and to suggest a synthesis of systems and processes that will contribute to mastery of the built environment. It is hoped that such a synthesis will lead to appropriate and effective mechanically coordinated low-energy architectural forms.

GARY R. STEFFY
IES, IALD

PART ONE

SENSORY PERFORMANCE STANDARDS: BASIC SPATIAL PATTERNS

Architecture reflects, in part, people's continuing attempts to establish a protected environment that approximates the sensory conditions with which they are most comfortable and at ease (in spite of the fact that these precise conditions appear only occasionally and unpredictably in nature). In addition to the historical building-shelter implications of this goal, many twentieth-century energy-consuming mechanical devices have been designed to fulfill this goal.

In beginning a study of environmental-control function of buildings or spaces as protection against sometimes hostile or disruptive external environmental conditions, an initial concern must be the character of the human sensory system. Human occupants perceive light as surface brightness and color; they absorb heat from warmer surfaces and warmer air; and they emit heat to cooler surfaces and cooler air. They respond physiologically to humidity, air motion, electromagnetic radiation, and air "freshness." They also respond to sound vibrations. In each case the intensity of response is influenced both by metabolic needs and subjective levels of adaptation.

A major function of a building is to accommodate all these sensory perceptions, to establish and maintain order in the sensory environment. This requires the designer to recognize and perhaps resolve problems stemming from three basic classifications of environmental influence of human behavior:

1. *Inhibiting environmental influences:* spatial conditions that produce distortion or disorientation or inhibit involvement, participation, performance of an intended activity or function, or user attitude and motivation

2. *Benign environmental influences:* neutral spatial influences that negligibly or variably influence user behavior and attitude

3. *Reinforcing environmental influences:* spatial conditions that may simplify user needs, such as orientation and spatial comprehension; reinforce or simplify performance of intended activities; or reinforce appropriate user attitudes, motivation, and social interaction

When considering matters of environmental "fit" and methods of achieving them, the reader should recognize that late twentieth-century architectural methods are clearly affected by a value system significantly different from that of previous decades of design. This modified value system is attributable to the evolving limits on the use of fossil-fuel-generated energy for operating building systems and to the increased substitution of renewable-energy-fueled systems. In this sense 1973 can be seen as a philosophical turning point in American architectural technology, particularly with regard to energy-consuming devices. While building-energy limits have become a matter of intense concern, the study of building design cannot, however, be limited to energy consumption alone: problems of human environmental needs remain. This leaves the designer with the theoretical and professional challenge of achieving high-quality environmental goals and specifications within the context of limited energy demand and reasonable building and equipment costs.

With this background in mind, the first section of this book focuses on the identification of appropriate performance specifications for the human sensory environment. It discusses the nature of the sensory background that supports human activities and behavioral patterns.

CHAPTER 1

The Luminous Environment

An important objective of this chapter is to demonstrate that light and dark are not antagonistic but rather complementary phenomena. Without shade, or dark, light loses much of its effectiveness as a communicating medium; prominence and rendition of details in the visual field are revealed by patterns of light and dark. In the overall setting, these light and dark patterns constitute environmental contrast. Through the careful manipulation of these light and dark patterns, designers can shape and modify the visual experience of a room. They do this by manipulating the perceptual role of lighting—both to facilitate visual tasks and to define the visual boundaries and hierarchy of a space, area, or activity. Designers also manipulate the psychological role of lighting to help establish a sense of intimacy, cheerfulness, or somberness that is appropriate as background for the intended activity. Designers manipulate the *functional role* of lighting as well. Extreme environmental contrasts can result in poor task perception and consequently poor task performance by occupants. Light intensities and surface finishes must be carefully coordinated.

This chapter opens with some of the basic aspects of vision and perception. Light will then be discussed in a structural sense—in terms of its ability to affect our emotional response to as well as our perception of space. As the chapter progresses, subjective and objective criteria will be introduced, in particular, visual impressions, brightness intensity and focal brightness, glare, color perception, visibility of task detail, illuminance, and power budgeting. These are the major factors comprising the performance requirements for the visual environment.

VISION

The sense of vision is based on the eye's ability to absorb and process selectively that portion of the electromagnetic spectrum that we call *light*. This sense is particularly vital, for it is used for most functions that require a grasp of spatial relationships and detail. Initially, vision includes the process of orientation and the formation of spatial impressions. Vision also involves scanning a variety of information cues, making simultaneous or successive comparisons, and assigning mental priorities regarding importance. And it involves communication—both the identification of meaningful information sources and the subsequent gaining of fine quantitative and qualitative information. Finally, vision is used for interpreting movement and rates of change.

Visual Sensation

The human eye responds to a very narrow portion of the electromagnetic spectrum. The region of 380 nanometers (deep blue) to 760 nanometers (deep red) generally defines the *visible spectrum* or *light*.

The eye is most responsive in the yellow-green region (550-560 nanometers); sensitivity diminishes toward deep blue at one end and deep red at the other (fig. 1-1). The eye is therefore essentially unresponsive to both ultraviolet and infrared wavelengths, which are immediately adjacent to the visible spectrum. (Excessive concentrations of infrared energy can, however, heat the cornea and lens and can cause damage. Similarly, excessive exposure of the eye to ultraviolet wavelengths below 310 nanometers can cause inflammation.)

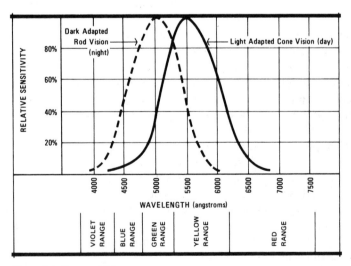

Figure 1-1. Eye sensitivity.

Maximum perception of fine detail—visual acuity—occurs when the image falls on the *fovea* (fig. 1-2). This is the central portion of the retina, predominantly made up of *cone* cells. Outside of the fovea, the concentration of cones diminishes, as normal vision depends primarily on the less responsive *rod* cells.

In general, visual response is dependent on the intensity of light (radiant energy or flux) that is incident on the eye and the time necessary for the flux to produce a sensation. If exposure time is inadequate, response is diminished—or even eliminated—accordingly. However, if time is adequate, response depends almost completely on incident flux, which can be translated into surface or object brightness. With very low brightness approaching *night vision* conditions, visual response tends to depend on the rod cells. As brightness on the retina increases, the more perceptive cone cells become increasingly active and acuity increases sharply until an optimum response condition is approached (see fig. 1-46).

Perception of Color

In addition to their contribution in the perception of fine detail, cone cells also exhibit good selective response to color. Rod vision, on the other hand, is generally crude vision that is deficient in both detail and color perception.

Color detail tends to be limited to the central foveal area of human vision; little color detail is perceptible in our peripheral vision where the rods are dominant.

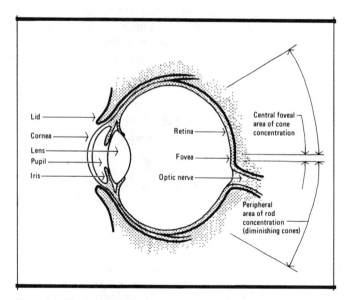

Figure 1-2. The human eye.

As noted previously, dependence on rod vision becomes increasingly dominant at very low intensity levels at which the cones lose their ability to respond. In this regard, note in figure 1-1 that peak eye sensitivity for *scotopic,* or rod vision occurs at a lower range—about 510 nanometers (dotted line). A result of this shift in sensitivity is the blue or blue-gray tone we experience with night vision.

Even under normal brightness intensities when the cones are fully operative, variations in the physiological response to color can be noted. For example, colors are not imaged identically on the retina of the eye. Some research studies report that when a green object is imaged clearly, a red object of the same size and in the same position would be imaged slightly larger and behind the retina. At the same time, a blue object would appear slightly smaller and in front of the retina. This optical relationship may help explain why *warm* colors appear to advance, while *cool* colors seem to recede (and diminish in size).

Accommodation

As the eye surveys an environment or a task it will, on intervals, fixate or focus. This process of focusing is called *accommodation.* Constantly changing focus or accommodating can lead to visual fatigue. On the other hand, constant, long-term focusing on a particular task also can result in visual fatigue. To minimize the possibility of the latter, visually distant focuses (e.g. window views, artwork, etc.) should be available. To minimize the possibility of the former, related task elements should be kept at equal distances from the observer. For example, a "hardcopy" paper document used as reference for a video display terminal (VDT) operator should be kept at the same plane and distance as the VDT screen. This may necessitate a paper document holder.

Response to Brightness

The total brightness range to which the eye is sensitive is from faint starlight to sunlight. This is a range of approximately $10^{10}:1$. However, at any single instant the brightness range is much more restricted. At any one moment the eye can readily distinguish a brightness range of approximately $100:1$ with good acuity; if moderate time is allowed for adaptation to either brighter or darker conditions, the eye can accurately distinguish a range of well over $1,000:1$.

Adaptation

The eye tends to seek a state of equilibrium that is appropriate for general brightness conditions. This constant adjustment involves some photochemical action, but it is most significantly affected by the action of the *iris,* which opens and closes to control the quantity of light that is permitted into the interior of the eye. The iris opens when the environmental brightness is low and closes to reduce light penetration when the environmental brightness is high. Since a time lag is involved in the expansion or contraction of the iris, the response of the retina in interpreting brightness at any given moment will be affected by the intensity of brightness in the general visual field during the immediately preceding period of time.

Judgment of Brightness Differences

In the final analysis, perception of brightness is a subjective phenomenon that varies with the adaptation of the individual observer. Even under optimum conditions, the eye is a very poor photometer. For example, when the human eye is called upon to estimate equal steps in brightness, it will seriously underestimate the difference. Normally, if subjects are shown a white surface (80 percent reflectance) and a black surface and then asked to select a gray finish that is approximately midway between these two, many will tend to select a 15-20 percent reflectance gray. These individuals may be quite surprised at the apparent lightness of the 40-45 percent gray tint that is the true photometric midpoint.

The ability to perceive minor brightness differences is also somewhat erratic and variable. When the occupant is visually adapted to high-intensity daylight-type conditions, it may be possible for him to perceive detailed brightness differences that vary from the average by as little as 1 percent. When lower environmental brightness conditions are involved, however, minimum perceptible differences must vary 5-10 percent (or more) from the average.

Glare

Visual response to brightness also depends on the distribution of light over the retina. When extremely unequal brightnesses are present in the visual field at the same time, the more extreme excitation of one part of the retina may inhibit the performance of other areas. When this occurs, perception of lower-intensity detail is seriously impeded by the brighter glare source. This condition is classically observed by drivers who are confronted with oncoming bright headlights on a darkened highway. A more typical and ever-present condition for the shopper or office worker is observed in spaces using the newer, compact low voltage incandescent accent lamps or where moderate-to-high wattage high intensity discharge lamps are used in downlights and in accent lights. The resultant overstimulation of a small portion of the retina makes perception of environmental detail extremely difficult.

Table 1-1. Typical effects of aging on ability to detect detail.

Approximate Age (years)	Relative Visibility (%)
20	100
30	95
40	87
50	74
60	59
70	35
80	21

SOURCE: Blackwell and Blackwell 1979.

Table 1-2. Typical effects of aging on glare sensitivity for typical interior environments.

Approximate Age (years)	Relative Sensitivity (%)
20	100
30	100
40	100
50	120
60	150
70	200
80	280

SOURCE: Blackwell and Blackwell 1979.

Effect of Aging

The physiological capabilities of the eye tend to deteriorate with age. There is a reduction in speed of perception, a reduced resistance to glare, and a lengthening in the time required for adaptation. Because of these factors, there is a measurable reduction in visibility and an increase in glare sensitivity (tables 1-1 and 1-2). This deterioration is particularly evident for night vision and in dimly lit environments.

The Visual Field

Normal human binocular vision involves a field of view approximately defined as 60 degrees upward from the line of sight, 70 degrees downward, and 180 degrees horizontally (fig. 1-3). However, acute perception of fine detail (foveal vision) takes place within an extremely small angle that paral-

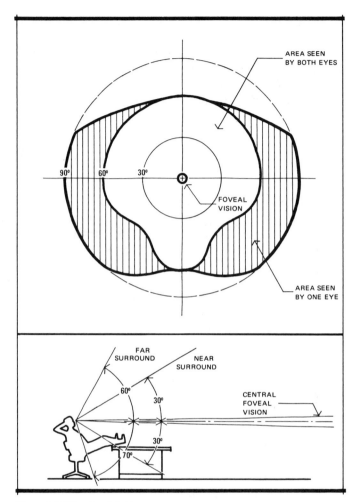

Figure 1-3. The visual field.

lels the line of sight. This angle is not greater than 2 degrees in diameter and is generally so small that the eye must change position in order to focus successively on both dots of a colon (:) viewed at a distance of 14 inches.

Outside of this small area of precise perception is a much broader area of peripheral vision in which perception of detail and color becomes successively less distinct. The more central portion of this peripheral vision is defined as a cone of approximately 30 degrees above, below, and to each side of center. This cone, the *near surround,* involves relatively clear imagery. A surrounding cone of general comprehension extends approximately 60 degrees above and to each side and 70 degrees down from the line of sight. Beyond this, an area extending approximately 85-90 degrees to either side of each eye registers major forms as indistinct masses.

Within the more peripheral portions of the field of view, the eye identifies information cues by responding to changes in brightness patterns and intensities, making this area relatively sensitive to movement, flicker, and the like. Peripheral vision is therefore more useful in enabling the viewer to maintain a sense of general orientation and a sense of relationship to the dynamic activities in the space.

Perception of Fine Detail in the Visual Field

In moderate and higher brightness environments (conditions in which cone vision predominates), maximum acuity occurs for detail that is located in the direct line of sight (0 degrees from the fovea). Acuity diminishes as the detail is moved to the periphery, so that about 15 degrees from the direct line of sight, acuity is approximately 15 percent of maximum. Under very low brightness conditions (conditions in which rod vision becomes dominant), visual acuity is about 10 percent of normal maximum, and acuity peaks for detail located about 4 degrees from the fovea.

Detection of Movement

While peripheral vision is highly useful for detecting movement or similar changes in the general background, perception of such changes remains most sensitive near the direct line of sight. Movement rates of approximately 0.9 minutes of arc per second ($\frac{1}{32}$ inch at a distance of 10 feet) can be detected in this central area, while movement must approximate 18 minutes of arc per second to be detectable in the more peripheral areas.

A similar pattern affects visual perception of detail in motion. Acuity of foveal vision for an object moving at a rate of 50 degrees of arc per second is approximately 60 percent of that achieved for a stationary object. When the moving object is located in the periphery, reductions in acuity of this magnitude occur for objects moving at much slower rates of speed.

The Luminous Environment

In general, then, human vision involves a narrow area of sharp central vision (foveal vision) and a much larger out-of-focus background (peripheral vision). Foveal vision includes the sensation of detail and color (cone vision), while the more inclusive peripheral vision (rod vision) is essentially responsive to brightness and mass. Within this context, background brightness patterns tend to become significant because they affect the general sense of spatial orientation that guides the occupant of a space.

To develop this further, background brightness patterns are important because foveal vision is guided or *cued* by information gained through peripheral vision. Potentially significant visual patterns are initially identified and located by scanning and assimilating the total visual field. Then central foveal vision is focused on the relevant detail that has been identified in the periphery. The ensuing study using foveal vision is the means by which the viewer derives most of the information necessary for specific orientation and the performance of a specific activity or task. But even while this detailed study is under way in the foveal area, peripheral vision is continually used to maintain general orientation and to identify any new information cues in the environment.

Design of the luminous environment is therefore primarily concerned with two aspects of human sensory behavior:

1. The visual task of *spatial orientation,* which requires the designer to be concerned with the effect of light in defining the space, the structural enclosure, or the activity, without introducing irrelevant patterns or visual confusion

2. Detailed *central task vision,* which requires the designer to be concerned with the effect of light in defining significant information centers and in assisting the accurate communication of visual detail required for acceptable performance of normal activities

The balanced manipulation of these visual conditions should provide for the viewer's need to judge distances and recognize relevant objects, materials, colors, and forms. At the same time, this environmental balance should reflect the need to protect the occupant from glare and from meaningless visual cues that may confuse his sense of orientation and purpose.

A viewer interprets the general environmental background largely through dominant brightness relationships. The subjective sensation of visual space is primarily a function of brightness pattern and pattern organization—the relationship of surfaces lighted or left in relative darkness.

THE FUNCTIONS OF ENVIRONMENTAL LIGHTING
Light Patterns

Light, of course, affects how well work tasks can be seen (visibility) and subsequently affects how well they are performed (productivity). This aspect of lighting is discussed later in this chapter. Of equal or even more importance, however, is how light affects the visual quality of a space and how it affects the sense of well-being experienced by users of that space. Light patterns influence our sense of space, our impressions, and consequently our actions. Therefore, the designer should become sensitive to the uses of sparkle, silhouette, focal emphasis, color tone, and other forms of spatial light. The designer must recognize that the correct use of light patterns is fundamental in satisfying some space-activity requirements, such as reinforcing attraction or attention, enhancing impressions of spaciousness, stimulating sensations of spatial intimacy or warmth, or reinforcing impressions of cheerfulness. The careful application of these principles combined with the principles of work-task lighting will ensure that the available lighting watts and available lighting equipment budget are used in the most beneficial and effective way.

Environmental Lighting as Part of a System of Spatial Cues

In recent years architectural researchers have been redefining functional design by investigating evidence suggesting that human responses to environmental lighting are, to some extent, shared experiences (Flynn 1977; Flynn and Spencer 1977; Flynn et al. 1979). This effort has led to the testing of the theory that some patterns of light might be "communicative" in the sense that they suggest or reinforce ideas shared by users having the same cultural or social background.

Individuals exchange ideas and information in many ways, and while human societies share much information through words in various spoken, written, and printed forms, some categories of information are communicated through symbolic visual patterns. Commercial trademarks are one example of this; railroad and traffic signals are another. Socially and culturally related individuals and groups obtain impressions of meaning by recognizing forms such as the Christian cross, the Star of David, and other symbols that identify cultural ideas, rituals, and groups. Lighting serves a similar symbolic function. Different forms of illumination—"intimate" lighting, the use of color, spotlighting, simulated daylight, and so on—provide culturally conditioned behavioral and affective clues guiding the user of an environment. Moreover, some visual forms provide a sense of spatial limits as well as scale reference, as figure 1-4 illustrates. Similar examples are white lane-lines painted along a road or an exposed modular ceiling grid which contrasts against the field (e.g., red grid against gray ceiling tile).

All of this indicates that comprehensive and sensitive lighting design must acknowledge and deal with higher-order lighting variables, recognizing the existence of patterns that communicate meaningful spatial information but are essentially independent of measured or calculated brightness quantities or other quantitative intensity specifications. As the designer changes lighting modes—patterns of highlight, shade, and color tone in a room—visual signals, forms, and cues will also change, altering users' impressions of meaning and importance. Thus lighting should be approached as a complex system of designed patterns that affect users' experience, spatial comprehension, and behavior within any built environment.

Figure 1-4. Spatial cues and spatial limits. The orderly array of light and dark contrasting elements produces a strong visual sense of territoriality.

Using Environmental Lighting to Provide Behavioral Cues

In terms of behavioral impact, there are two basic categories of environmental lighting systems. The first category consists of lighting systems that flood a space somewhat indiscriminately. These systems tend to be behaviorally neutral in the sense that they tend *not* to exert an intentional reinforcing or guiding influence on user impressions or behavior. They may offer advantages in space utility, flexibility, and general clarity. For some types of activities, however, the resulting diffuse and uniform light distributions are rather significant shortcomings because of the bland psychological effect created in the room. Figures 1-5 and 1-6 illustrate typical effects of this method of illumination when little thought is given to the three-dimensional aspects of space design. In situations requiring this flood of light, careful use of surface color will help to minimize blandness, as well as to establish environmental communication. Articulated ceiling systems such as vaulted modules or modular, accented suspension elements (for example double-T grid systems with black reveals) may also help establish visual interest.

The second category consists of lighting systems that develop specific patterns of light and shade to reinforce selected information or room cues. More arousing behaviorally, this type of environmental lighting is intended to reinforce a specific pattern of user impression or behavior and thus requires more specific design intentions. This approach is based on the idea

Figure 1-5. Indiscriminate lighting — lost visual cues.
This uniform incandescent downlighting system does little to enhance the visual environment.
Note the miscued wall area to the right of the artwork — the lighted wall attracts attention,
leaving the artwork as a secondary focus in relative shadow.

Figure 1-6. Indiscriminate lighting — low contrast.
The combination of a uniform fluorescent lighting system with medium reflectance neutral
finishes results in a low-contrast, uninteresting visual environment.

that light can influence a user's selective attention or alter the information content of the visual field. These lighting designs should be carefully evaluated for their role in establishing *cues* that reinforce the user's understanding of the environment and surrounding activities. In this regard, the following categories of visual experience should be considered.

The Effect of Light on User Orientation and Room Comprehension

Some lighting patterns seem to affect personal orientation and user understanding of the room and its artifacts. For example, spot-lighting or high-contrast focal lighting, as shown in figures 1-7a and b, affects user attention

Figure 1-7a. Specific light patterning—room function comprehension.
Focal lighting of architectural details and niches against a relatively dark background
produces a sense of drama in this gallery/corridor. Room function comprehension is less
likely to be a "long corridor" but more likely a visually interesting architectural envelope
punctuated with art niches (see Figure 1-7b).

Figure 1-7b.

Table 1-3. Brightness ratios.

Visual Impact	Focus-to-Background Brightness Ratio
Barely recognizable contrast; negligible attraction power as a focal point (see fig. 1-6)	2:1
Minimum meaningful contrast as a focal point; marginal attraction power	10:1
Dominating contrast as a focal point; strong attraction power (see figs. 1-7a. and b.)	approaching 100:1

SOURCE: Flynn 1978.

Figure 1-8. Specific light patterning—room size comprehension.
Highlighting the walls in this space helps produce a comprehension of a wider-than-actual room size.

and consciousness. Table 1-3 indicates the visual impact of various brightness ratios. Wall lighting, as shown in figure 1-8, and corner lighting affect user understanding of room size and shape, establishing or modifying the sense of visual limits or enclosure.

The Effect of Light on Impressions of Activity, Setting, or Mood

Other lighting patterns seem to involve the communication of ideas or impressions, suggesting that light is, in part, a medium that assists communication of spatial ideas and moods. In this sense lighting patterns can assist the designer in creating impressions of somberness, playfulness, pleasantness, tension, and other qualities. The designer can also use light patterns to affect psychosocial impressions such as intimacy, privacy, and warmth. Lighting can be used to produce a carnivallike atmosphere or to produce an austere atmosphere for meditation. Lighting can produce a cold, impersonal public place or reinforce an impression of a warm, intimate place where a greater sense of privacy is felt. These impressions or moods are often fundamental in satisfying experience and activity requirements in a designed space. Lighting to enhance these impressions should be recognized as more than aesthetics; it is a tool for influencing human behavior, performance, and productivity.

LIGHT-STRUCTURE MODELS

In an era of limited energy resources when lighting watts must be scrutinized with professional attention to function and value, designers must be very selective in the use of lighting. There are three brightness parameters which seem to influence specific subjective impressions of observers. These are brightness uniformity (i.e., uniform vs. nonuniform), brightness position (i.e., overhead vs. peripheral), and brightness intensity (i.e., bright vs. dim). Research indicates that these brightness parameters or *modes* can be orchestrated or *structured* to achieve specific subjective impressions of visual clarity, spaciousness, relaxation, and privacy. The light-structure models in figures 1-9 through 1-12 have been developed to serve as partial guides for the use of environmental lighting effects appropriate for various task and nontask applications (Flynn 1977). These figures illustrate how the three brightness modes (uniformity, location, and intensity) interact to elicit various degrees of each of the four subjective impressions.

The research work in this area is not complete. For example, the three lighting modes are based on relative scales. No quantitative data has been estimated. Nevertheless, these modes can be used to assist the designer in setting significant design directions in the early phases of a project.

Uniformity indicates the range of brightness variations necessary to elicit a given impression (e.g., "uniform" means little or no range in brightness variations; the lighting system should provide uniform brightness over the entire location mode). Location indicates the primary room zone at which the suggested "uniformity" and "intensity" should occur to elicit a given impression. Intensity indicates the relative level of brightness necessary to elicit a given impression.

Impressions of Visual Clarity

See figure 1-9. Visual clarity, a subjective visual impression of luminous environments, refers to the perceived distinctiveness of architectural and

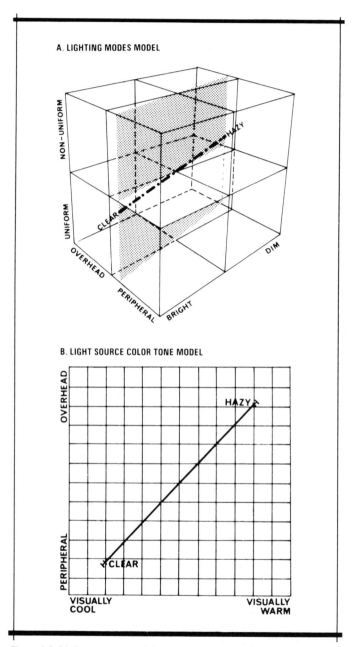

Figure 1-9. Light structure model: impressions of visual clarity.

human features and details. The term *clear* is used to signify a luminous environment that promotes impressions of distinct visual perception of a space and the objects within, whereas the term *hazy* signifies a luminous environment that promotes impressions of somewhat clouded or fuzzy visual perception of a space and the objects within. Visual clarity is:

- an important subjective factor to be considered in the design of work spaces
- a subjective visual impression that appears to be reinforced by three lighting influences:
  1. uniformity (reinforced somewhat by more uniform brightness)
  2. location (reinforced somewhat by more peripheral brightness)
  3. intensity (reinforced by relatively high brightness)

Impressions of Spaciousness

See figure 1-10. Spaciousness refers to the perceived size of the architectural enclosure around an individual. *Large* signifies a luminous environment that promotes impressions of expanded spatial limits—increased volume. *Small* signifies a luminous environment that promotes impressions of confinement. Spaciousness is:

- an important subjective factor to be considered in the design of circulation and assembly spaces, such as corridors, lobbies, and assembly halls
- a subjective visual impression that appears to be reinforced by three lighting influences:
  1. uniformity (reinforced by uniform brightness)
  2. location (reinforced by peripheral brightness, e.g., wall lighting)
  3. intensity (reinforced somewhat by relatively high brightness)

Color of the light itself—warm or cool—appears to be a negligible factor. Recognize, however, that cool-colored surfaces appear to recede, while warm-colored surfaces appear to advance.

Impressions of Relaxation

See figure 1-11. Relaxation refers to the perceived intentions of human function. *Tense* signifies a luminous environment that promotes impressions of fast-paced visual work. *Relaxed* on the other hand signifies a luminous environment that promotes impressions of rather comfortably paced activities, including visual work. Relaxation is:

- an important subjective factor to be considered in the design of more casual areas, such as waiting rooms, lounges, some restaurants, and conference areas

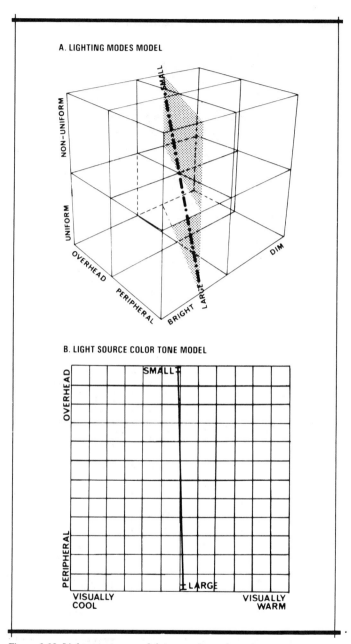

Figure 1-10. Light structure model: impressions of spaciousness.

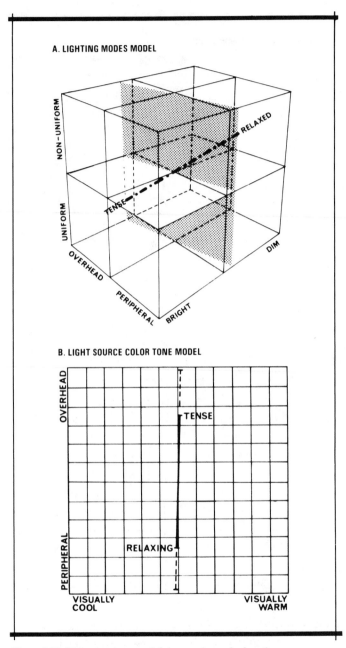

Figure 1-11. Light structure model: impressions of relaxation.

• a subjective visual impression that appears to be reinforced by the following lighting influences:
 1. uniformity (reinforced by rather nonuniform brightnesses)
 2. location (reinforced by peripheral brightness)
 3. intensity (reinforced somewhat by relatively lower brightness)

Impressions of Privacy

See figure 1-12. Privacy refers to the perceived intentions of spatial use. *Public* signifies a luminous environment that promotes impressions of extroverted and deliberate activity. *Private* signifies a luminous environment that promotes impressions of introverted and subdued activity. Privacy is:

• an important subjective factor to be considered in the design of more intimate casual areas, such as nightclubs, some restaurants, and residential living spaces
• a subjective visual impression that appears to be reinforced by the following lighting influences:
 1. uniformity (reinforced by rather nonuniform brightnesses)
 2. location (reinforced somewhat by peripheral brightness)
 3. intensity (reinforced by relatively lower brightness)

Individuals derive successive patterns of information by scanning the boundaries of a space, activity, or task, thereby assembling a conception of spatial limits and relative position and direction. The light-structure principles can assist this information-scanning process. Occupant participation in the space is best with minimal visual interference or distraction from the environment. Therefore, orderly application of light-structure principles is necessary. Figure 1-13 exemplifies the use of the visual clarity light-structure model.

SPATIAL ORDER AND FORM
Visual "Noise"

Clutter in the visual environment is analogous to noise in the acoustical environment. Visual noise is a particularly significant factor when intensive visual attention is required, such as in a library. Although the ability to process visual cues depends first of all on an individual's concentration, visual sensory overload occurs when the mind attempts to process too many visual cues. In such circumstances the individual's ability to perform a given function may decrease, as indicated in figure 1-14. Therefore, in spaces

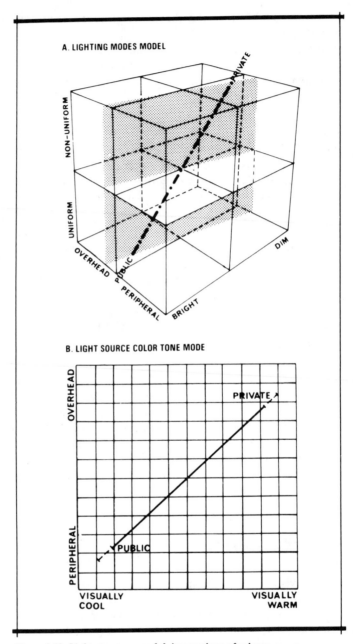

Figure 1-12. Light structure model: impressions of privacy.

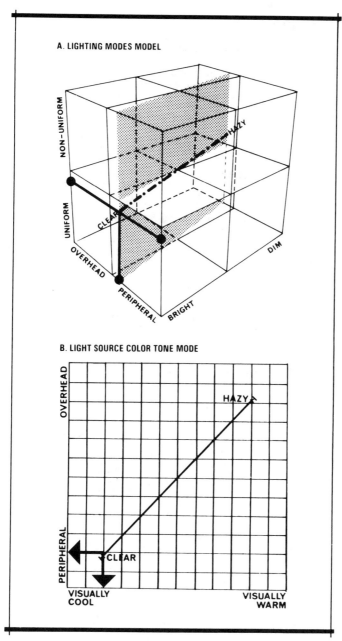

A. LIGHTING MODES MODEL

B. LIGHT SOURCE COLOR TONE MODE

Figure 1-13. Light structure model: impressions of visual clarity. An example of the use of the model. (continued)

EXAMPLE OF MODELS AS DESIGN GUIDES

STEP 1: ESTABLISH SPATIAL FUNCTION:
Secretarial Pool

STEP 2: ESTABLISH DESIRED VISUAL CLARITY IMPRESSION:
Clear

STEP 3: UNIFORM VS. NON-UNIFORM MODE (SEE LIGHTING MODES MODEL):
Extending a line from the Clear/Hazy Plane, and perpendicular to the Non-Uniform Plane, there is an indication that a somewhat uniform lighting mode will help promote a "clear" impression.

STEP 4: OVERHEAD VS. PERIPHERAL MODE (SEE LIGHTING MODES MODEL):
Extending a line along the Clear/Hazy Plane, and perpendicular to the Overhead/Peripheral Plane, there is an indication that a lighting mode which tends to be near the architectural periphery will help to promote a "clear" impression.

STEP 5: BRIGHT VS. DIM MODE (SEE LIGHTING MODES MODEL):
Extending a line from the Clear/Hazy Plane, and perpendicular to the Bright/Dim Plane, there is an indication that a bright lighting mode will promote a "clear" impression.

STEP 6: COOL VS. WARM (SEE LIGHT SOURCE COLOR TONE MODEL:
There is an indication that cool colored light sources in the architectural periphery help promote "clear" impressions. This is found by extending a line from the Clear/Hazy axis and perpendicular to the Cool/Warm axis, as well as extending a line perpendicularly to the Peripheral/Overhead axis.

STEP 7: LIGHTING DESIGN SUGGESTIONS:
The previous exercise suggests that in order to promote impressions of visual clarity, "clear", the lighting system should tend to be rather uniform with some peripheral, cool colored light, and should tend to be rather bright. Note that this brightness can (and should) be achieved not only with fairly significant quantitites of light, but also with high surface reflectances.

Figure 1-13 (continued). Light structure model: impressions of visual clarity.

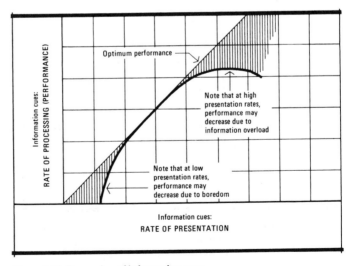

Figure 1-14. Processing of information cues.

where complex tasks are performed, the visual environment should be simplified by minimizing irrelevant or meaningless cues. Figure 1-15 is an example of unusually dominant ceiling articulation produced by the luminaire pattern. In this case the lighting system is responsible for the "visually noisy" aspect of the library. Luminaire and brightness patterns should be developed to simplify the process of orientation and spatial definition.The lighting designer has the responsibility to design the lighting system around the three-dimensional qualities and the task aspects of a space. "Plan design" (2-dimensional planning) of lighting systems can lead to visual cluttering, as in the ceiling shown in figure 1-15. This lighting system does not respond to the visual needs of the occupants.

Visual sensory deprivation can be as serious a problem as visual sensory overload. As shown in figure 1-14, low information presentation rates can result in a decrease in performance caused by boredom. The environment shown earlier in figure 1-6 is one of minimal visual stimulation, where boredom may result unless appropriate colors are used throughout the space.

Figure 1-16 shows a lighting design in which particular attention was given to the location and prominence of the lighting system. Notice that although *light* is used to articulate the space and activities, luminaires do not introduce confusing patterns in the ceiling nor contribute to meaningless spatial cues, as might a more convenient system of track lighting.

Figure 1-15. Spatial order—information cues—visual noise.

Figure 1-16. Spatial order—3-dimensional space/function articulation.

Luminaires

In developing a sense of visual order it is important to separate the illumination function of the lighting system from the problem of luminaire placement. While luminaires are often placed overhead (on, in, or suspended from the ceiling), very often the objects that we use for visual orientation are located in the lower part of the visual field. At the same time, the eye is drawn to bright areas (such as luminaires) that contrast with the general background condition. Accordingly, if the luminaires are to be a relevant part of the visual background, the brightness pattern must merge with and reinforce the spatial patterns that are meaningful to the activity.

Since the luminaires in the room may become a significant determinant of visual form, the layout of these luminaires should establish a sense of scale, visual order, and appropriate direction (if any). An activity-oriented location of luminaires may contribute to a sense of spatial organization when the focal points and activity areas remain stable, as shown in figure 1-17. When activities are subject to change and the space must be versatile, however, moderate to low brightness luminaire patterns may merge with the ceiling plane in such a way that they become a unified part of this major architectural form.

Indirect lighting must be carefully analyzed for not only luminaire patterns, but also brightness patterns. Although some brightness patterning on

Figure 1-17. Spatial order—activity-oriented lighting.

the ceiling may be desirable for "visual interest," the ceiling plane in office settings should maintain a reasonably constant brightness so that the ceiling is perceived as a unified architectural form.

Whether the lighting is direct or indirect, the layout of luminaires can reinforce the sense of direction and enhance spatial perspective. This is most critical in long, narrow corridors, where the "tunnel effect" can be avoided by appropriate luminaire placement and subsequent brightness patterning (figure 1-18).

Lighted Surfaces

As illustrated in figures 1-16 and 1-19, the lighting system can delineate the form and character of major surfaces that define the space or function. The form of the light distribution should enhance the affected surface. Lighting can also be used to define the intersection of major surfaces, further clarifying architectural space and orientation. This is shown in figure 1-20. Miscuing is generated when irrelevant light patterns are caused by the inappropriate selection of light distribution. If surfaces are to be perceived as integrated architectural forms, then odd light scallops and irregular brightness patterns should be avoided.

This is not to imply that irregular light patterns are never desirable. Depending on the intended spatial use, desired visual impression and architectural intent, irregular brightnesses may add visual interest and depth.

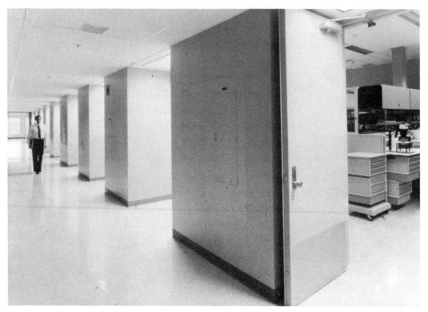

Figure 1-18. Brightness patterning—reinforcing room entries and de-emphasizing corridor length.

Figure 1-19. Brightness patterning—form delineation.

Figure 1-20. Brightness patterning — edge delineation.

These are, however, novel effects that serve as temporary visual stimulants and may be considered as "visual noise" in work-task environments.

BRIGHTNESS
Spatial Distribution of Brightness

The development of information content in the visual field may vary with the behavioral needs of the activity. In this sense the background can be developed either to emphasize or subordinate various aspects of the environment or activity. As one example, a system can be designed to produce horizontal illumination over the normal task or circulation plane, while subordinating the vertical and overhead elements (fig. 1-21). The more intensively illuminated central area causes people and activities in the lower visual field to become the dominant brightness feature. The enclosure itself appears as a neutral or even subordinate visual influence. This situation will tend to increase the occupant's normal awareness of nearby detail, of other people, and their movement. In a relative sense, this condition seems to encourage an attitude of social involvement among the occupants, thus providing a public impression.

When visual emphasis is focused on vertical and overhead architectural elements through brightness relationships while reducing lower horizontal surface brightness (fig. 1-22), foreground detail is lost as objects and people

Figure 1-21. Brightness patterning—horizontal emphasis.

Figure 1-22. Brightness patterning—vertical emphasis.

Table 1-4. Brightness ratios affecting public–private impressions.

Impression	Horizontal-to-Vertical-Background Brightness Ratios
Tendency toward a social behavioral pattern and active movement— impression of public space	1 : 1 to 100 : 1 (horizontal is brighter)
Tendency toward an introspective behavioral pattern and more relaxed movement—impression of privacy (see *figure 1-24*)	1 : 20 to 1 : 100 (background verticals are brighter)

in the central area tend to go into silhouette. This condition makes the foreground activities visually subordinate and tends to induce a more intro- spective attitude in the occupant—creating an intimate atmosphere in which the individual feels a sense of privacy or anonymity. Table 1-4 indicates the brightness ratios between foreground horizontal surfaces and background vertical surfaces necessary to affect impressions.

General Intensity of Spatial Brightness

The subjective sensation attributable to architectural spaces is affected by the intensity of brightness, as well as the previously discussed organization of lighted areas and surfaces. Although the basic methods of introducing light into a space may remain constant, significant variations in the general level of brightness will affect the number and intensity of interreflections of light between surfaces in the space. This interreflected light has a diffusion influence, altering subjective and objective qualities in the space.

At the low end of the brightness scale, for example, the luminous *glow* induced by a dimmed light setting limits the diffusion of a multidirectional lighting system, and creates sharp contrasts in a room (figs. 1-23, 1-24, and 1-25). Higher light settings, on the other hand, increase the intensity of interreflections, reducing shadow and silhouette and creating subdued con- trasts due more to the color and finish of surfaces (fig. 1-26). Furthermore, a slight increase in intensity at the low end of the brightness scale will produce a significant improvement in the individual's ability to discriminate detail and color. As brightness increases the *rate* of improvement diminishes and the environment approaches a condition of delivering occupants' maximum acuity regarding the spatial background (the surrounding environment). High general brightness intensities tend to contribute to an impression of increased activity and efficiency, while lower brightness intensities tend to reinforce impressions of slower-paced activity—relaxation and privacy.

Figure 1-23. Brightness intensity — low setting, high contrast (bright focal element, dim surround).

Figure 1-24. Brightness mode — peripheral, dim (privacy, relaxation).

Figure 1-25. Brightness intensity — low setting, high contrast (bright side walls, dark ceiling, table, and end wall).

Figure 1-26. Brightness intensity — high setting, low contrast.

Excessive Brightness and Glare

While brightness and brightness contrast are basic in visual communication, excessive contrast or excessive background brightness can disrupt the ability of the eye to perceive fine detail. In the extreme, these *glare* conditions can temporarily cripple vision by destroying the observer's ability to perceive a task, an obstruction, an object, or a space adequately.

The experience of driving toward a late afternoon sun, the after-dark effect of approaching bright headlights, or the occasionally overpowering brightness caused by the sun on clean sand or on white snow are common experiences that reveal the nature of *disability glare*. This is glare of sufficient intensity to impair visual acuity and to impair the occupant's ability to orient himself.

A somewhat more subtle form of glare is experienced with unshielded fluorescent lamps, excessive luminaire brightness or, today, with relatively high wattage compact fluorescent and HID downlights. Although the resulting disability tends to be relatively moderate, temporary physical disability again results due to unequal excitation of the retina (see previous discussion of *Glare* under *Visual Sensation*).

Glare is generally corrected by reducing the source luminance (such as by dimming or by delamping to fewer lamps per luminaire or lower wattage/light output per lamp); by the use of baffles, louvers, or diffusers to reduce excessive luminaire brightness; by source relocation outside of the normal visual field; and by reducing the reflectance characteristics of excessively bright surfaces.

Brightness Tolerance as a Function of Area

In estimating and evaluating brightness tolerance, there is a fundamental relationship between *brightness intensity* and *area of brightness*. This relationship will affect the actual quantitative limits of visual comfort. Table 1-5 lists typical diffusing materials used for lighting along with corresponding allowable lamp lumen limits.

Brightness Tolerance as a Function of Location

As a corollary to the relationship between intensity and area of brightness, the negative influence of a glare source depends upon its location in the normal field of view and its proximity to the central foveal area of the eye. Figure 1-27 illustrates typical average brightness levels that can be tolerated in different portions of the peripheral field of view. (*Maximum* brightness of relatively small highlight areas can be as high as three times the *average* brightness shown.) While these tolerances will vary somewhat with the state of adaptation of the occupant's eye, this diagram indicates that there must be

Table 1-5. Limits of visual comfort (area).

Diffuser Material	Lamp Density (Luminous Ceiling): Maximum Allowable Generated Lumens/sq ft of Diffuser Surface
White diffusing plastic, flat	600
White diffusing plastic, formed	700
Clear prismatic plastic	1150
Small-cell plastic louvers	925
Brushed aluminum louvers	1150
White enameled louvers	1400
Gray enameled louvers	2300
Black enameled louvers	3500

For *uniform* patterns of luminaires that involve less than the full luminous ceilings indicated above, the allowable generated lumens/sq ft can be greater than that shown in the table.
Multiplier factors (for the above generated lumens/sq ft):

100% Ceiling coverage	1.0
10% Ceiling coverage	4.0
1% Ceiling coverage	8.0
0.1% Ceiling coverage	16.0

NOTE: Because of their dominating influence in the visual field, luminous walls involve more critical brightness limits. Limit generated lumens/sq ft to 60-75 percent of the values indicated above for ceiling areas.

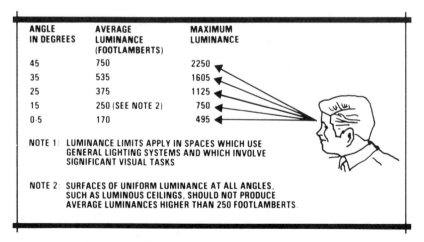

Figure 1-27. Luminance limits of luminaires for visual comfort.

increasing restriction of general brightness as the area in question approaches the center of the visual field.

These studies help to explain why brightness levels that are considered acceptable for luminous ceilings are found to be excessive and uncomfortable for luminous wall areas. Wall areas must function within more restrictive tolerances because they represent a more dominant influence in the normal visual field. This is a critical issue. Vertical surfaces serve a significant function in the visual interpretation of space, yet their illumination must be done with extreme care, giving attention to surface reflectance, color, and illumination intensity.

Adaptation and Surround Brightness

The subjective impression of visual comfort also depends on the brightness relationship between task surfaces and their surroundings. Facing a window with a view of a bright overcast sky can make reading a book extremely difficult because of the effects of background glare. Equally difficult is reading a brightly illuminated book when the surroundings are in darkness.

In spaces where sustained visual work is involved (such as offices, classrooms, and industrial areas), brightness relationships within the normal field of view should be controlled to allow the eye to adapt to an overall environmental brightness near the brightness of the task itself. In this way the *shock effect* of high environmental contrast as well as the strain of continual readaptation can be minimized.

In areas designed for prolonged work, therefore, research and experience have indicated the need for lighting the ceiling and walls (as well as the working surfaces) to avoid uncomfortable or fatiguing working conditions produced by excessive contrast. For comfortable seeing over a long period of time, the general brightness of surfaces immediately surrounding the task should not differ appreciably from that of the task itself. For work areas, it is generally recommended that spatial brightnesses average no less than one tenth and no more than ten times the average brightness of the task.

Visual Comfort

Visual comfort performance standards are therefore important to overall user satisfaction within a given space. Along with the average and maximum luminance limits in figure 1-27, a single-number comfort rating has been developed for standard lensed and louvered luminaires in a monolithic ceiling plane. This rating, known as Visual Comfort Probability (VCP), is an indication of the percent of population that may be expected to find a lighting system acceptable from a glare standpoint (Kaufman 1981). For instance, a VCP of 70 (maximum equals 100) indicates that the given lighting system in a space does not produce a discomfort glare situation for at least 70 percent of the occupants. It should be recognized that this Visual Comfort Probability indicator does not provide any information regarding how comfortable the occupants may be under the given lighting system. This measure determines only the percentage of population that will be under the threshold for discomfort glare.

In the past, some indirect systems have been heralded as having VCP ratings of 100. This is a fallacy, as ceiling finishes and luminaire-to-ceiling proximity greatly affect glare conditions for indirect systems. Indirect lighting systems as well as vaulted and coffered ceiling systems do not fall within the original VCP system of visual comfort rating. Research may lead to visual comfort prediction techniques for this equipment, and the designer should review current literature.

A mockup of luminaires side by side can be of great assistance in establishing a luminaire's glare potential by simulating all "angles of view." It provides a more accurate view than data generated in a photometric test report, which is generally limited to five degree or ten degree increments (see chapter 5). These increments are usually not small enough to catch very high luminaire brightnesses at specific odd-angles.

Ceiling finish can affect luminaire glare conditions. The darker a ceiling, the greater the chances of luminaire glare being a problem. Acoustical baffles and ceiling "beams" can help minimize luminaire glare conditions, as can regressing luminaire lenses into the luminaire housing.

Sparkle

If glare is an undesirable element in the environment, then the difference between *glare* and *sparkle* is an important design consideration. The principal difference lies in the relationship between brightness intensity and area, and in the prominence of that area in relation to the total visual field. If large areas of brightness are distracting and disconcerting to the viewer, relatively small areas of similar (or higher) intensity may be the points of sparkle and highlight that contribute visual interest and a sense of spatial vitality.

While the negative influence of glare is to be minimized, then, this should generally be done without eliminating the closely associated stimulating influence of sparkle.

PERCEPTION OF DETAILED VISUAL INFORMATION

As implied in the previous discussion of the luminous background, the spatial lighting pattern should contribute to the identification of significant information centers—including task and work centers. But as soon as this identification has taken place for a given occupant, the suitability of the lighting on the task area itself will depend on its quality in assisting communication of precise visual detail. The remainder of this chapter will therefore focus on the effect of light on the specific ability to communicate a visual idea or message.

For purposes of this discussion, *communication* is defined as the act through which people derive the information necessary to gain knowledge to make an effective decision, or to perform a meaningful task. This process involves the use of the visual sense for observing situations and relationships, or for discriminating and differentiating precise detail. In this sense, the designer is concerned with the general context within which the communication is perceived; with the identification of general meaning; and with the perception of meaningful differences.

Effect of Spatial Context

Brightness in the peripheral areas surrounding a specific, localized task center has an important effect on the ability to distinguish fine task detail (visual acuity). Optimum acuity is achieved when the general brightness difference between the central task (foveal vision) and the immediate spatial background (peripheral vision) is from 1 : 1 to 4 : 1, with the task area tending to be slightly brighter than the background. An increase in this ratio to 250 : 1 (task is brighter) will produce a reduction in acuity of approximately 10 percent; while for a bright task seen against a totally dark background, acuity is reduced approximately 20 percent.

As a general rule, when highly precise visual performance is required, spatial brightness differences exceeding 10 : 1 should be kept well outside of the more central 40 degree visual cone. (But of course this restriction is not generally applicable for more casual visual activities where the drama of high contrast focal centers can make an important contribution in the *experience* of the space or activity.)

Even more significant differences in acuity occur when the spatial background is brighter than the task (when the task detail approaches a silhouette condition). A relatively moderate 1 : 20 ratio (background is brighter) will produce a reduction in acuity of approximately 20 percent—and these reductions multiply rapidly as the background intensity increases. Consider the reduced visibility of detail in the silhouetted chairs in figure 1-24, or in the detail of the table and centerpiece in figure 1-25.

Dark work surfaces seen against bright spatial backgrounds should be avoided where precise perception of detail is required for effective visual performance and/or prolonged participation. (On the other hand, spatial silhouette may be useful when it is desired that the occupant be principally aware of general forms and spatial context while deemphasizing his awareness of foreground detail. This is the so-called *cocktail lounge effect* that contributes to a sense of personal detachment or privacy. See previous discussions, *Spatial Distribution of Brightness* and *Light Structure Models: Impressions of Privacy,* see fig. 1-24.)

COLOR

Subtle changes in the color tone of light (whiteness) can influence the subconscious judgment of the general environment. Perceptual awareness of this aspect of light is most intense when a change first occurs or when the individual first enters a space, before the eye has time to adapt to the new color condition.

Different color tones of light may suggest the pinks and purples of a sunrise or a sunset, evoke the impression of warm sunlight or cool overcast skies, or produce a completely unnatural visual situation. Each of these conditions can be produced within the broad category of white light, although each is actually deficient in some portion of the spectrum. A shift in spectral emphasis is most immediately noticeable in the perception of neutral surfaces and familiar surface tones such as human skin tones, but it affects the perception of all surface colors and color differences, graying some colors while increasing the relative vividness of others.

Subtle shifts in the perception of surface tones and colors will affect the sense of warmth or coolness associated with the visual space. We tend to associate a warm visual atmosphere with hues of yellow through orange and red to red-purple. Warm light sources like the sun, many incandescent lamps, and the older warm white fluorescent lamps tend to create a dominant impression of visual warmth by emphasizing these hues while graying others. On the other hand, cool light sources, such as skylight and some fluorescent, mercury vapor, and metal halide lamps emphasize the colors that tend to create a cool visual atmosphere, from hues of blue-purple through blue and blue-green to yellow-green.

Ongoing research in this area may lead to definitive guidelines on the color atmosphere and its impact on sensory responses. Recent research indicates, for instance, that the color tone of an environment may affect perception of environmental temperature (Rohles 1977). Eventually, designers may be in a position to conserve or more effectively use energy through careful analysis of visually modified sensory responses.

Perception of Color and Chromatic Contrast

Colored materials reflect or transmit light wavelengths selectively in certain regions of the spectrum, and the color that is perceived in an object or surface is determined by the spectral characteristics of this transmitted or reflected light. This interdependence of light and color means that in order to provide accurate color rendition the light source must emit those wavelengths that the object is able to reflect (or transmit). A deficient mixture will alter perception and cause the impression that specific colors are deficient or completely lacking. For example, a green object under a red light source appears black or dark gray because the surface absorbs all colors except green, and little or no green is present in the red light to be reflected. Any spectral deficiencies that are inherent in the prevailing light source will cause some surface colors to be *grayed*. This action tends to affect contrast adversely and therefore reduces acuity. Classic examples of deficient light sources include standard high pressure sodium—a yellow light is emitted rendering many surfaces and materials a muddy gray. The light source selected should therefore produce energy in those regions of the spectrum that are meaningful for the task or the decorative scheme. Figure 1-28 graphically depicts the interdependence of light-source transmission characteristics and surface-reflectance characteristics, and resultant perception.

Detection of Color Differences

Differences in the spectral quality of white light can exert a decisive influence on acuity when the detection of subtle differences in surface color is involved (such as in color matching or color inspection). In most cases, white light that is rich in the spectral region where maximum *absorption* (minimum reflectance) occurs will tend to accentuate narrow differences in surface color (table 1-6).

Apart from precise inspection situations, these principles also underscore the important relationship between color of light and surface color when specifying fabrics, wall colors, and other elements in the spatial scheme. In the final specification, colored materials should be appraised, matched,

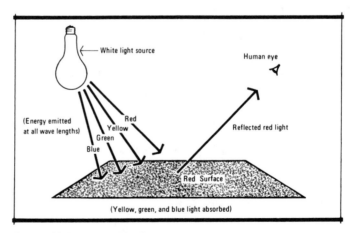

Figure 1-28. Perception of color.

Table 1-6. Detection of differences when matching colors.

Desired Viewing Condition	Suggested Light Source Spectral Characteristics
Optimum detection of differences in blues, purples, blue-greens	Warm-toned source that is rich in red, yellow
Optimum detection of differences in reds, yellows, yellow-greens	Cool-toned source that is rich in blue, green

Table 1-7. Color (vector influence).

Color of Light	Relative Luminance Required for Equal Attraction
White light	1.0
Yellow light	1.2
Red light	0.3
Green light	0.4
Blue light	0.6

and selected under the lighting conditions that will predominate in the actual space.

Effect of Color Accents

A vector influence can be produced with color (chromatic contrast) as well as with luminance contrast. Colors of light exhibit varying attraction values that are completely independent of brightness (table 1-7).

Although the results of color accent studies are somewhat variable, they indicate that saturated colors of red, green, or blue light will compete for attention with white light of greater intensity. Yellow light must be slightly brighter than white and considerably brighter than the other colors for equal attraction value.

Appearance of Human Complexions

The spectral characteristics of the dominant light source are similarly significant in altering the appearance of human complexions. White sources that are rich in energy at the red end of the spectrum (incandescent and tri-phosphor fluorescent) complement and flatter the complexion, imparting a ruddy or tanned character to the skin. White light sources that are strong in the yellow and blue ranges of the spectrum but weak in red (mercury, standard metal halide, and cool white fluorescent) tend to produce a sallow or pale appearance.

Color Rendering

A single-number system has been developed to indicate the ability of light sources to render colors (Kaufman 1981). Although a Color Rendering Index (CRI) of 100 is considered "best," and although many incandescent lamps have CRI's ranging from 95 to 100, a CRI less than 100 does not mean that a lamp is "poor." It does mean, however, that it is likely that some colors may be rendered better or worse than by a reference incandescent lamp. This "better or worse" ambiguity acts as a warning to the designer to review or select finishes and colors under the actual design-intended source. Further CRI can be used as a selection criteria only when comparing lamps of the same lamp color or color temperature (see below).

Lamp Color

Up to this point, the discussion has centered around the colors of surfaces and subsequent visual impressions. The lamps and/or filtered or lensed luminaires themselves, when energized, produce an inherent color. Except for some of the new very high-efficiency lamps, this inherent color is broadly classified as "white." This ranges, however, from a very cool white to a very warm white.

This color of light, or lamp color, is known as color temperature. Measured in degrees Kelvin, it is essentially the temperature an iron ingot would need to be to produce the same surface color as the lamp surface color when the lamp is energized. Color temperature may affect the subjective interpre-

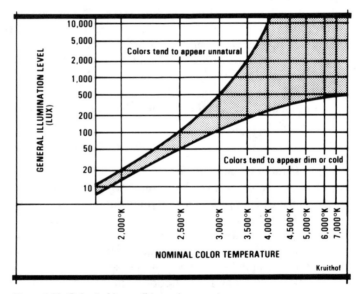

Figure 1-29. Color "whiteness" (amenity curve).

tation of brightness intensity. Cooler toned light sources of same intensity as warmer toned counterparts may seem to appear brighter. Research by Kruithof (1941) led to development of the Kruithof Amenity Curve shown in figure 1-29. There is very little substantive basis, however, for this curve and for the "cooler is brighter" generalization. Kruithof's work was based on very few subjects who may have understood the premise of the research prior to their participation. The amenity curve is no longer considered an appropriate guide to design decisions.

For design purposes, typically lamp color temperatures are maintained consistently (matched) throughout a project, unless a color differentiation is desired.

VISUAL PERFORMANCE POTENTIAL: VISIBILITY

Consider for a moment all of the environmental factors that affect work performances in an office. The previous discussions have dealt with the surrounding visual environment—with performance related to color and luminances of the visual surround. Other factors affecting worker performance are the acoustical and thermal environments. Perhaps, though, the single most critical factor affecting worker performance is the worker's own ability to perform—motor control, visual functioning, memory, and so on. The best that a team of designers can do is to provide the *potential* for high visual performance. (For purposes of this discussion, it is assumed that all other environmental factors have been appropriately handled and that only visibility remains a factor.)

Visibility refers simply to "how well an individual can see the task." A task that cannot be seen very well is said to have *poor visibility.* In certain instances, this may be a negligible factor. For example, guest registration at an art exhibit opening is not terribly important, and even though instructions may request last names first, it is of little or no consequence if guests cannot readily read and follow this instruction. On the other hand, a bank customer filling out a loan application must be able to read the instructions easily and see where information is to be written on the form. Good visibility is obviously important to the successful completion of the loan application.

Experimental evidence links visibility with worker performance. Ongoing research in this area may soon produce definitive guidelines regarding visibility performance standards and desired productivity levels for certain tasks. An understanding of some basic principles and of the visibility metrics can assist today's designer in providing good visibility in work environments.

Task Characteristics

Before spending time, effort, and money on computer calculations and/or full-scale mockups of "solutions," the designer must fully understand and evaluate the physical displays involved in "task performance." This can lead

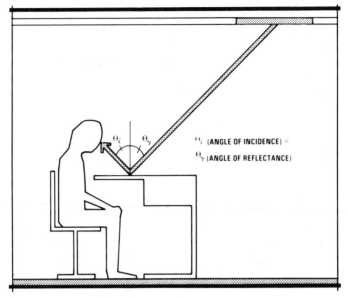

Figure 1-30. Physical phenomena: reflectance angle equals incidence angle.

to a mutual effort by designer and client to provide maximum visual performance potential—visibility—without necessarily resorting to expensive lighting equipment and/or power intensive lighting systems. For example, a simple maintenance procedure involving frequent replacement of the ribbon in computer printout equipment may lead to clearer output copies and therefore a more visible task display. This "solution" for task visibility is less energy intensive and less costly than designing a new lighting system to provide very high illuminance levels. Some task displays can cause decreased visibility if they are not properly illuminated. On specular or "shiny" materials, light follows the principle of "angle of reflectance equals angle of incidence" as illustrated in figure 1-30. Although this phenomena is clearly evident in mirrored surfaces, it is also obvious in glossy magazines, and to a lesser extent in many pencil or ink tasks (ballpoint pen, press-printed material, xerography) if the light source and the observer's eyes are geometrically in the wrong spatial relationship.

Task Contrast and Luminance Contrast

Direct luminaires are said to be located in the *offending zone* when they are located in positions where light reflects harshly from a task display or portions of a task display into the eyes. The offending zone is illustrated in figures 1-31 and 1-32. A glossy brochure task is shown in figure 1-33, with

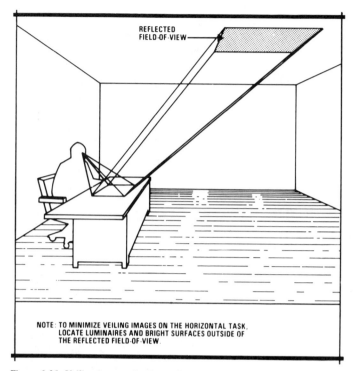

Figure 1-31. Veiling images (horizontal surfaces).

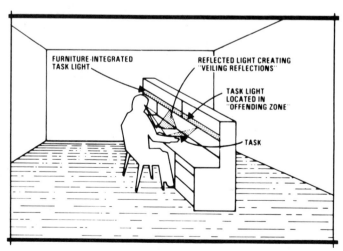

Figure 1-32. Offending zone: furniture-integrated standard lens or bare lamp lighting/normal task orientation.

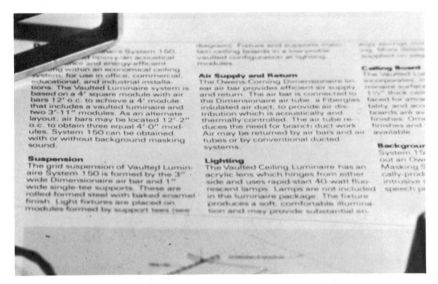

Figure 1-33. High visual performance potential: luminaire to side of task produces good task visibility.

illumination provided from a direct luminaire located to the side of the visual display. Notice the high contrast, a good reading condition. When the direct luminaire is moved into the offending zone, a reduction in visibility results (fig. 1-34) because the letters on the paper are reflecting as much light to the eye as the paper itself—hence no contrast. The light reflections resulting in visibility reduction, or no contrast, are known as *veiling reflections,* as they tend to "veil" the task from the observer, making task performance difficult if not impossible.

Most surfaces, materials, and objects do not reflect light uniformly in all directions (diffusely). Depending on how light strikes a surface and the inherent reflectance characteristics of the surface, more or less light is reflected. This type of contrast is known as *luminance contrast,* which can be defined in the following manner:

$$\text{luminance contrast} = \frac{\text{background luminance} - \text{detail luminance}}{\text{background luminance}}$$

Background luminance refers to the measured brightness of the general background of a specific task. For example, this page has a white background, and the luminance of this white background can be measured. Detail luminance refers to the measured brightness of the portion of the task that conveys meaningful information. For example, the luminance of the black letters on this page can be measured and constitutes detail luminance.

Figure 1-34. Low visual performance potential: luminaire in offending zone produces poor task visibility.

The higher the difference between background luminance and detail luminance, the higher the luminance contrast, the higher the task visibility.

Luminance is also known as *exitance,* that is, the amount of light radiating or exiting in a specified direction from a surface. In figure 1-34, background exitance and detail exitance in the direction of the "eye" (camera) are equal; hence their difference is zero, hence luminance contrast is zero.

Maximizing luminance contrast maximizes visibility. The designer, therefore, must be familiar with the visual tasks to be performed within spaces in order to design a lighting system that maximizes luminance contrast.

Visibility Performance Standards

One objective of the lighting designer is to provide maximum visual performance potential in interior work environments. Over the last fifteen years, a variety of metrics have been developed to measure visual performance potential. Since this area of research is still in a state of flux, no conclusive metric is considered a standard. Recent research has identified contrast alone as a valid visibility performance standard (Rea 1981). Since contrast can be predicted and measured for a variety of standard office tasks, this metric should prove quite useful in developing visual environments that provide the potential for good visual performance. The lighting designer is encouraged to review current periodicals and literature as more information regarding visibility and productivity becomes available.

Minimizing Veiling Reflections

There are a variety of methods for providing high task visibility. The designer is urged to perform cost-benefit analyses, since these methods may or may not be cost effective depending on the importance of the task, task reflectance characteristics, task size, and desired visual performance rates. As figure 1-33 illustrates, luminaires strategically located away from the offending zone can provide high task visibility. Note that the angles involved will change with the orientation of the task. For instance, a drafting task on a table tilted 25 degrees will have a different area defined as the "offending zone" than that illustrated in figure 1-31. The offending zone for a vertical task orientation is shown in figure 1-35.

Recognize that veiling reflections are a result of a significant amount of light coming from a specific direction and subsequently reflecting in a specific direction. If directionality of the incoming light can be minimized, then veiling reflections can be minimized. Therefore using more direct luminaires of less intensity per luminaire is helpful. Or, using indirect lighting which, if used in white-ceiling spaces can produce multidirectional, diffuse lighting.

In open-plan office arrangements, a ceiling-mounted luminaire will undoubtedly be in the offending zone at some point in time as the furniture is rearranged. Polarizing panels in ceiling-mounted luminaires can help mini-

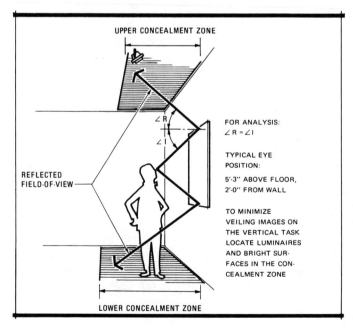

Figure 1-35. Offending zone: vertical task orientation.

mize veiling reflections. As discussed in chapter 2, however, lenses provide hard reflective surfaces from which sound can reflect into one workspace from another. Therefore furniture-integrated lighting equipment may seem appropriate. Many times the only place for the furniture-integrated lighting equipment is in the offending zone (figure 1-32). For these cases, specially designed batwing-distribution lenses, polarizing lenses or specially designed photometric lamp sleeves are recommended for minimizing veiling reflections.

Finally, by increasing illuminance onto a task, visibility may often be increased. This approach, which generally means increasing energy consumption, was popular prior to the 1973 oil embargo. In part, this explains the sometimes very high illuminance levels prescribed in the 1950s and 1960s. Except where all other efforts fail to increase visibility, it is no longer practical to resort to this solution for minimizing veiling reflections.

Luminance Contrast and the Perception of Form and Texture

Three-dimensional form is *seen* as a relationship of highlight and shadow. When this relationship is altered (by changing the directional characteristics of the lighting system), the impression of form may also be changed. In this sense, the lighting system should perform in sympathy with the inherent surface character of materials that define the physical space.

Grazing light will emphasize highlight and shadow (see fig. 1-36). As a factor in visual acuity, this condition will improve perception of depth— which, in turn, may be important in the perception of texture or in the

Figure 1-36. Surface detail.

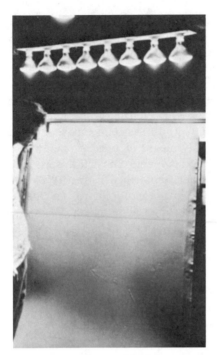 

Figure 1-37. Surface blemishes: grazing light.   *Figure 1-38. Surface blemishes: frontal light.*

accurate performance of a very precise three-dimensional visual task. For example, grazing light will often aid in the perception of surface blemishes and errors in finish workmanship (see figs. 1-37 and 1-38). This then is a visual task that is representative of those that depend on direction of light rather than intensity or color to produce appropriate conditions for visual perception.

Conversely, a more frontal lighting condition will reduce visibility of surface flaws and create a visual impression of flatness or surface uniformity that is more suitable for a plaster wall or tile ceiling. The diffusing effect may be the result of illumination emitted directly from luminaires, and/or it may be produced by interreflection from other high reflectance surfaces (such as floor, walls, or ceiling). In either case, the action of *frontal* light is particularly significant in reducing or eliminating the localized and random brightness variations that are due to subtle irregularities and flaws.

When the central visual task involves the perception of texture and three-dimensional form, this perception is influenced decisively by the character and distribution of light that impinges on the task.

The illustrations in figure 1-39 suggest the considerable range of visual impressions that result from changes in the directional composition of the light. They also suggest that analysis of *visual form* involves more than the physical form itself. Visual form is physical form as modeled by light.

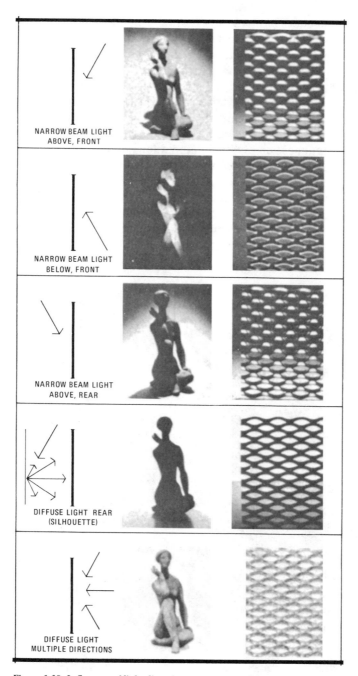

NARROW BEAM LIGHT
ABOVE, FRONT

NARROW BEAM LIGHT
BELOW, FRONT

NARROW BEAM LIGHT
ABOVE, REAR

DIFFUSE LIGHT REAR
(SILHOUETTE)

DIFFUSE LIGHT
MULTIPLE DIRECTIONS

Figure 1-39. Influence of light direction on texture and form.

Negative Influences of Shadow in Central Communication Areas

For many common visual tasks, shadow may become an element of distraction in the immediate task center. In some cases, shadows on the work surface may be mildly irritating (such as those produced by the hand while attempting to write under a concentrating light source).

Where more precise and demanding visual study is involved, shadows produced by this same concentrating lighting condition may become extremely disconcerting—and in some cases, hazardous in the sense that the shadows impede communication of visual information necessary for adequate safety (see figs. 1-40 and 1-41). The excessive concentration and constant readaptation required of a worker in these situations can, over a period of sustained work, result in visual fatigue, accidents, and errors.

However, this does not mean that highlight and shadow are never desirable in intensive working environments. Previous reference has been made to the advantages of grazing light in those instances that involve perception of

Figure 1-40. Environmental shadow in the work area.

Figure 1-41. Environmental diffusion in the work area.

three-dimensional tasks. Furthermore, just as highlight and shadow on a sunny day become emotionally stimulating visual influences, carefully placed brightness accents and shadow areas are useful for visual relief and interest in the interior environment. However, in most cases a diffuse condition is desirable at the task center itself, and the effect of environmental shadows should be minimized in this area.

Detail Size

Most persons with normal vision can distinguish black detail on a white background if the detail subtends at least one minute (1′) of arc at the center of the eye. At a distance of 100 feet this indicates a typical minimum detail dimension of approximately 0.4 inches; at 1,000 feet, a detailed dimension of about 4 inches is minimum. Where closer tasks are involved, a minimum of about 0.004 inch is the approximate minimum when the viewing distance is 12 inches.

The Relationship Between Brightness and Size of Detail

Optimum acuity is more easily achieved when the detail size exceeds this typical threshold condition. Increased detail size may facilitate relatively good visual communication even under somewhat adverse conditions. An example is the experience of relatively easy reading of a newspaper headline in a dimly lighted room; while careful reading of smaller type under the same illumination becomes much more difficult and prolonged.

The nature of this relationship between size of detail and brightness is suggested in figure 1-42. The indicated nomograph relationship exists for well-defined detail and form seen against varying reflectance backgrounds.

Recognition and Memory

Experience and memory are also factors in visual communication. Some forms are well known to the observer and will be recognized under relatively poor visual conditions. This recognition is possible because the observer need not study the precise detail in order for a message or idea to be communicated to him. Commercial trademarks and traffic sign shapes are among the more obvious examples of this. In a similar way, quick recognition of many numbers, letter forms, and other common shapes depends on this principle.

Questions of size and contrast must therefore be evaluated in terms of the task. Is simple recognition satisfactory (as in pleasure reading)? Or is it necessary to study precise detail (as in many office and industrial tasks)?

Similarly, the character and quality of the light must be evaluated for its effect on visual recognition. Time for perception may be prolonged if the

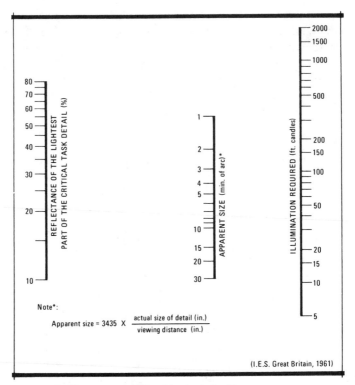

Figure 1-42. Illumination relationships (task detail).

lighting is not *typical* or if the lighting does not provide the expected, remembered, or learned perception. For example, the act of searching for a red automobile can be prolonged under a mercury vapor or high pressure sodium light source that is deficient in red energy.

Conditioned Association

For the most part, this discussion of communication has been concerned with perception—and in this sense it may be relevant to make some distinction between *perception* and *aesthetics.*

Communication of the *meaning* of what we see is based on *perception.* Is it round? Is it flat? Is it an orange? Is it an apple? Judgment or *value* is based on *aesthetics.* Is it a good orange? Is it a pleasant texture? Is it an appropriate space?

Often these value judgments are based on previous experience—on memory of the meaning of a similar perception or idea. Some of these associations are related to the nature of the light that impinges on the detail, and in this sense the lighting system may exert a direct influence on aesthetic judgment.

Figure 1-43. Light from a natural direction.

Figure 1-44. Light from an unnatural direction.

Artists have long been aware of these variations in our perception of light and have skillfully utilized the psychological associations that these variations induce. In painting, for example, the contrast between *yellowish* highlights (from direct sunlight) and *bluish* shadow (illuminated by skylight) is sometimes exaggerated to achieve a *natural* impression.

In a similar way, a lighting condition that alters or reverses the usual direction of light changes the normal (natural) relationship of highlight and shadow. Aesthetic judgment may therefore be based on an unnatural perception, possibly inducing a sense of uncertainty, mystery, or even fear (figs. 1-43 and 1-44). While most situations probably favor the development of more natural environmental situations, special conditions can be visualized where the environmental condition is manipulated to produce a complementary aesthetic impression (such as the sense of mystery produced by lighting a Buddha or a Gothic gargoyle).

In cases that involve perception and judgment of objects and forms, then, variations in light (color of light as well as direction) will affect the observer's unconscious judgment of what he is seeing. Again, judgment is based on perception of the physical form *as modified by light.* But remembered associations may be decisive in the communication of an idea or experience.

Brightness of Typical Tasks

Research indicates that increasing light energy (flux) is generally required to maintain a constant level of acuity when the following conditions occur:

- As the size of the detail is reduced
- As the contrast between the detail and the background is reduced

- As the task reflectance is reduced
- As the time permitted or utilized for perception is reduced

As a generality, then, (and assuming constant contrast, size, and time for viewing) visual acuity increases with brightness. There is a particularly high rate of improvement when low initial intensities are involved. This generally reflects the increasing influence of *cone vision* over *rod-dominated vision* as brightness intensities increase from minimum conditions. Once the cones begin to approach full stimulation, acuity continues to improve as brightness increases, but at a significantly diminished rate (see fig. 1-45).

The Influence of Illuminance

For many of the often-encountered "standard" visual tasks, the background and the task reflectances are well known. In such cases, illuminance criteria are generally substituted for visibility (luminance-contrast) criteria. Although this provides for quicker design solutions, it carries with it the risk of the designer losing sight of luminance ratios and the possibility of overlooking nonstandard tasks and their special reflectance characteristics. In the not-too-distant future, luminance criteria may supplant illuminance critera.

The Illuminating Engineering Society of North America (IESNA or IES) has made an effort to emphasize the effect of luminance more heavily in the latest IES Illuminance Selection Procedure (Kaufman 1987). As task reflectance decreases, the task illuminance criteria may increase incrementally. Similarly, as observer age increases and subsequently observer visibility

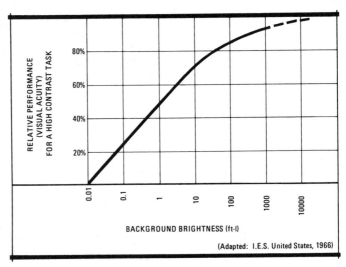

Figure 1-45. Illumination relationships (visual acuity).

decreases (see table 1-2), task illuminance criteria may increase. Finally, as the visual tasks become more important (critically important in the case of a head nurse reading patient status reports and doctors' instructions) and/or visual tasks must be performed faster (quality control on an assembly line), illuminance criteria may increase, responding to the visual acuity versus background luminance curve shown in figure 1-45. Illuminance guidelines for representative tasks are indicated in table 1-8. These guidelines are meant to apply to the *plane* of the task and the *area* in which the task is most likely to occur.

Task-Ambient Lighting

The term *task-ambient lighting* became a common part of the design vernacular in the 1970s. Unfortunately, many equate task-ambient to furniture-integrated. Since *task lighting* means "lighting the task" and *ambient lighting* means "lighting the surround," then it is rather obvious that the means to these ends can be many. Some solutions can provide both task and ambient illuminance. For instance, in a densely populated office, ceiling-integrated direct-indirect lighting oriented to each task location may provide proper task and ambient illuminance.

POWER AND ENERGY BUDGETING

New concepts in lighting equipment along with new understandings of human visual perception and visual productivity since the early 1970s can be used to provide attractive, productive, and energy-effective environments. Unfortunately, legislation has heavily emphasized *power* budgeting (watts per square foot), rather than *energy* budgeting (watts per hour or kilowatts per hour). This single-number criterion has resulted in lighting systems for many projects that do little else but boast record-breaking lows for lighting power. Hopefully, as concern grows for the human occupants of our buildings, this will take a more or less equal place with other criteria in performance specifications for the visual environment.

Lighting power refers to the inherent electrical requirement of the lighting system in order to produce light; while lighting energy refers to the use of that power over a period of time. To a certain extent, as we reduce lighting power of a given lighting system we also reduce task illuminance, reducing background luminance and subsequently reducing visual acuity, as shown in figure 1-45. Therefore, tasks may be performed at a somewhat slowed rate. Thus what may be saved in "electrical" dollars is perhaps lost in "human" dollars.

Recognize that careful use of lighting energy can result in high task visibility and can account for important occupant impressions (as discussed

Table 1-8. Typical task illuminance range guidelines (in footcandles).

	Range Guidelines					
Tasks	5-20	20-50	50-100	100-200	200-500	500-1000
General						
Circulation						
Corridors, escalators, elevators	●					
Lobbies	●					
Locker-, toilet-, and washrooms	●					
Storage						
Inactive	●					
Active (large items)	●					
Active (small items)		●				
Office						
Office work						
Reading, transcribing, filing			●			
Accounting, auditing, tabulating, business machine operation			●			
Cartography, design, drafting				●		
Word-processing video display terminal (VDT)	●					
Conference						
Conferring		●				
School						
Classrooms						
Classroom work, library, study		●				
Manual arts, drafting, laboratories			●			
Sewing				●		
Assembly						
Auditoriums, cafeteria, gymnasiums	●					
Indoor exhibition sports		●				

Store

Service areas

Merchandising areas

Showcases, displays

Industry

Inspection

General

Difficult

Very difficult

Assembly

Medium

Fine

Woodworking

Sizing, planing, rough sanding,
medium machine and bench work, gluing

Fine sanding and finishing

Printing

Sorting, font assembly

Machine composition, composing

Proofing, routing, macking, finishing,
tint-laying

Proofreading, color inspection

Packaging

Wrapping, packing, labeling

Food Service

Food selection

Display, cashier

Dining

Casual

Intimate

in *Light Structures,* earlier). Using these lighting criteria can lead to getting more for the energy dollar than may be possible using the power budget criteria alone. Remember, it is not how many watts of power, but how those watts are used that is important. New light sources are being introduced that are much more efficient than the previous standard fluorescent lamps, that is, the new lamps provide more light output per power input. Those new sources which are HID, however, have severe switching limitations due to warm-up time. For this reason these sources are generally zone switched. Unlike most of the office furniture-integrated and some of the office ceiling-integrated fluorescent lighting that is locally switchable, these new lamps must remain energized even when some office workers are not present. For this reason the new HID low-power-budget systems may consume as much energy as carefully designed locally switchable fluorescent systems. A recent trend is toward more compact sources. This provides improved light control and efficiency. These lamps, in fluorescent and high intensity discharge, are some of the most efficient white light sources available today.

The designer should study occupants' work habits, "task time," and clients' energy rates, and produce a life-cycle cost analysis of the various potential lighting solutions, rather than select the lowest power (lowest watts-per-square foot) system.

Metric Summary: Lighting

English Units		Metric Units
1 Footcandle (fc)	=	10.76 Lux (lumens per square meter)
1 Footlambert (fL)	=	3.426 Nits (candelas per square meter)
1 Foot	=	0.3048 Meters

REFERENCES

Blackwell, O. M., and H. R. Blackwell. 1979. Individual responses to lighting parameters for a population of 234 observers of varying ages. Report to the Illuminating Engineering Research Institute, New York, N.Y.

Flynn, J. E. 1977. A study of subjective responses to low energy and non-uniform lighting systems. *Lighting Design and Application* 7(2):6-14.

Flynn, J. E. 1978. The effect of spatial lighting on user behavior: A preliminary summary of theories and hypotheses to be considered. Unpublished report to Illuminating Engineering Research Institute, New York, N.Y. (IERI Project 92), February.

Flynn, J. E.; C. Hendrick; T. Spencer; and O. Martyniuk. 1979. A guide to methodology procedures for measuring subjective impressions in lighting. *Journal of the IES* 8: 95-107.

Flynn, J. E., and T. J. Spencer. 1977. The effects of light source color on user impression and satisfaction. *Journal of the IES* 6 (3):167-79.

Kaufman, J. E., ed. 1984. *Illuminating Engineering Society Handbook: 1984 Application Volume.* New York: IES.

Kaufman, J. E., ed. 1987. *Illuminating Engineering Society Handbook: 1987 Reference Volume.* New York: IES.

Rea, M. S. 1981. Visual performance with realistic methods of changing contrast. *Journal of the IES* 10 (April):164-77.

Rohles, F. H. 1977. The role of environmental antecedents in subsequent thermal comfort. ASHRAE (2449).

Steffy, G. R. 1990. *Architectural Lighting Design.* New York: Van Nostrand Reinhold.

CHAPTER 2

The Sonic Environment

HEARING

The sense of hearing is based on the ability of the human ear to process selectively a range of airborne vibrations. This sense is primarily used for communication of information through such devices as verbal dialogue or warning signals. As with vision, simultaneous or successive information cues are processed and assigned mental priorities regarding importance.

Hearing is often the decisive sense for precise communication of ideas between individuals and is particularly dominant when visual conditions are poor. Hearing may also affect spatial orientation in that there will likely be an instinctive desire to identify and locate the source of sonic signals. Furthermore, there may be a subjective sensation of spatial *hardness* or *softness* communicated through the sense of hearing.

The human ear detects sound over an extremely broad range of intensities and frequencies by responding to small variations in air pressure. These variations induce a subtle vibration of the ear drum; this in turn induces variations within the liquid-filled cochlear canal, where the auditory receptors are located (fig. 2-1). In terms of time separation, two clicks as near in sequence as 0.001 second can be detected by the ear as separate signals.

The ear does not, however, perceive the whole range of vibration frequencies as sound. For example, a tuning fork vibrating at a rate of 15 Hertz (Hz = cycles per second) arouses no sensation of hearing. Most people do not perceive sound until approximately 30 Hz. is reached.

The full range of human response to sound involves a frequency *spectrum* that ranges from approximately 30 Hz. at the low end to approximately 10,000 Hz. at the upper end. For young, healthy individuals not constantly exposed to very loud music, the range of sensitivity extends as

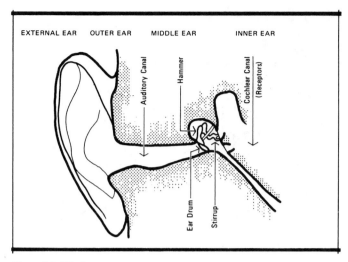

EXTERNAL EAR   OUTER EAR   MIDDLE EAR   INNER EAR

Auditory Canal

Hammer

Cochlear Canal (Receptors)

Ear Drum

Stirrup

Figure 2-1. The human ear.

low as 20 or as high as 20,000 Hz. Optimum sensitivity occurs in the mid-frequency range between 500 and 6,000 Hz. This range includes most of the normal *speech range.*

The Sonic Spectrum and the Area of Audible Sensation

The intensity of sound is measured in *decibels* (dB), and the listener's acuity (ability to hear detail) can be measured in terms of the threshold decibel levels at which the person can detect sound signals at various frequencies. The variable subjective sensation of sound intensity is summarized for pure tones in the form of *equal loudness contours* (fig. 2-2). These curves generally define the area of audible sensation. The 1,000 Hz. tone is the base intensity for each of these curves, and the curves indicate the intensity levels at each frequency that will be subjectively interpreted by a normal listener to be the same loudness as the 1,000 Hz. tone.

The Threshold of Hearing

Zero dB on the sound intensity scale is, by definition, the threshold of hearing at 1,000 Hz. Most normal listeners can detect a slightly lower intensity signal at 4,000 Hz; while considerably higher intensity levels are required to produce minimum perception at both ends of the frequency spectrum.

The threshold that is defined in this manner assumes an idealized *quiet* background. When background noise is present, however, a somewhat higher threshold is created. For example, while a 10-dB signal at 1,000 Hz. would be

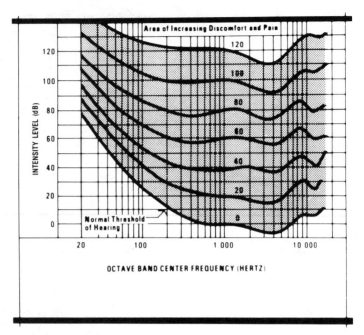

Figure 2-2. Ear sensitivity (equal loudness contours).

easily perceptible for normal human listeners in a quiet environment, a background noise level of 25 dB would establish another threshold that would eliminate perception of the original 10-dB sound. In order to be perceived by the same listeners, meaningful signals must now exceed the 25-dB background.

Distortion of High Intensity Signals

When background noise is very intense, it becomes quite difficult to communicate an intelligible signal. This is due in part to the high-level threshold. But is is also due to a condition within the ear that is somewhat analogous to *glare* in the eye. In this regard, efforts to increase the intensity of communicating signals (by shouting or by using electronic amplifiers) must be limited because a very high intensity signal will tend to produce interior distortions within the ear itself. These distortions will, in turn, obscure the clarity of the signal by introducing an internal masking effect.

When low or moderate background noises exist, then, improved communication can be facilitated by increasing the signal intensity. But when high levels of background noise are present in the environment the effectiveness of this technique is limited, and an improved communication condition will more likely depend on methods that reduce the higher threshold created by the noise.

Pain and Hearing Loss

The action of the hammer in the ear provides some muscular adjustment of the ear drum to prevent injury at high intensity levels. Exposure to sound pressure levels in excess of 120 dB (see top curve of figure 2-2) will generally cause increasing discomfort and pain in the ear. Such high intensities can produce momentary hearing disability or even permanent sensory damage. This intensity level 120 dB defines the upper limit of audible sensation, and is known as the *threshold of feeling* (discomfort and pain) (Yerges 1969).

An abrupt but less intense noise (such as the noise produced by a large exploding firecracker) will produce momentary deafness. This is somewhat similar to the effect caused by a higher background noise level over a short period of time, and as such is known as temporary threshold shift or TTS, since an individual's sensitivity to sound has been momentarily shifted from base threshold. Although individuals differ considerably in their vulnerability to such effects, the intensity and length of deafness will generally depend on the intensity and duration of the infringing high-intensity sound. Continuous or repeated exposure to moderately high noise levels can lead to permanent hearing loss. For example, when exposure to intensities exceeds the durations listed in table 2-2, permanent losses in sonic sensitivity can result. The values in table 2-2 are weighted averages across a range of frequencies, and as such are reported in "dBA"—dB refers to sound intensities at specific frequencies, while dBA refers to a weighted average of sound intensities over a range of frequencies as measured by a dBA meter.

The hearing loss that is associated with normal aging is termed *presbycusis*. The degree of hearing loss varies with frequency, the loss becoming more severe at the higher frequencies (Peterson and Gross 1974). This limits speech intelligibility by older individuals, since many consonant sounds are in the higher frequency ranges.

Multichannel Listening

The two ears can be used together when a single dominant signal is involved. To some extent they can be used as separate channels when several sources or signals are heard at the same time. If the information rates—the rates at which information is audibly signalled—are not too high, separation of sound sources can be easily accomplished by the listener without need of external modification of the signal itself. As the complexity and multiplicity of signals become greater, however, the listener will need to rely on one or more of three methods of separation:

1. Developing separation in amplitude (generally by moving closer to one sound source)
2. Developing separation in frequency (such as a conscious or subconscious decision to listen only to the lower frequency voice and disregard the higher frequency sources)
3. Developing a separation in space (possibly by turning the head so that one source is dominant on one ear, while sources acting on the other ear are ignored)

Spatial Judgments

While sonic spatial judgments are crude, multi-channel listening provides a limited auditory perspective that facilitates judgment of distance and direction. Judgment of *distance* is relatively simple, although not very accurate or precise. This may involve loudness (volume) and may also involve the complexity of the sound. In the latter case, the full-frequency modulation of a nearby sound becomes a simpler and more limited sonic spectrum as distance increases.

Directional judgments are of two types: those that require one ear and those that require two. For example, some judgment of sonic direction may be subconsciously based on previous knowledge, such as the association of traffic noise with the direction of a nearby road. Similarly, visual cues can take precedence over sonic cues, with the result that loudspeaker sounds from good systems are easily associated in direction with a movie screen or a lecturer. These examples represent readily localized sound sources and require only one ear for adequate hearing (or two ears working together). When previous knowledge and vision are both eliminated, however, a sense of direction is developed by using each ear separately. Generally, this is accomplished by movement of the head so that each ear is stimulated somewhat differently (fig. 2-3).

In summary, the spatial patterns that are involved in human hearing are generally based on sonic contrast. If a meaningful sound falls within the normal response range but is obscure due to a lack of intensity or lack of contrast with the background, listening becomes difficult or seemingly irrelevant. On the other hand, the sensitivity of the ear is limited in the sense that meaningful signals can also become too loud for pleasant and comfortable listening. Both of these extremes introduce adverse influences that can impede effective participation in an activity. Excessive or irrelevant background noise may also obstruct an activity by introducing an element of

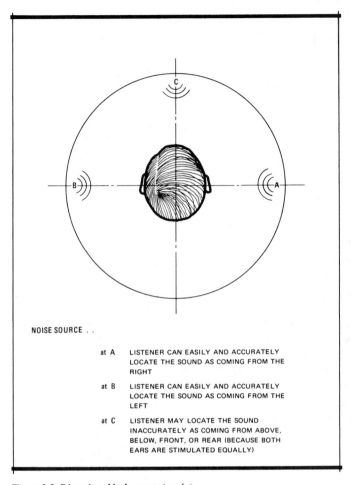

NOISE SOURCE . .

at A LISTENER CAN EASILY AND ACCURATELY LOCATE THE SOUND AS COMING FROM THE RIGHT

at B LISTENER CAN EASILY AND ACCURATELY LOCATE THE SOUND AS COMING FROM THE LEFT

at C LISTENER MAY LOCATE THE SOUND INACCURATELY AS COMING FROM ABOVE, BELOW, FRONT, OR REAR (BECAUSE BOTH EARS ARE STIMULATED EQUALLY)

Figure 2-3. Directional judgments (sonic).

distraction. The intensity of this distracting influence will depend on the tolerance and sensitivity of the individual listener; this may also vary for the same individual at different times.

In a still more nebulous area, sonic judgments may relate directly to the individual's orientation ability and ability to comprehend the nature of the environment. Hearing is used to make crude judgments of distance and direction when vision is lacking or obscured. Hearing also assists in perceiving a *sense of hardness* or *softness* that may be associated with a room.

Design of the sonic environment is therefore primarily concerned with two aspects of control: the *communicating signal* and the *sonic background.* The balanced manipulation of these sonic conditions should provide for the listener's need to receive communicated ideas by distinguishing relevant signals as separate from the general background. At the same time, this environmental balance should reflect the need to protect the individual occupant from excessive intensity levels and from irrelevant signals that may confuse the sense of orientation or distract attention from the primary activity.

PERCEPTION OF THE SONIC BACKGROUND

Subjective sensation of the sonic background is primarily a function of reverberation and random noise intensity. Both of these have some influence on the occupant's sense of spatial appropriateness and well-being. Both will also influence the ability to transmit and receive a sonic communication.

Reverberation

When a sound source ceases to emit energy, the direct sound stops immediately. The energy within an enclosed space, however, will continue to reflect between the room surfaces; as these waves successively pass the listener's ear, the original sound will continue to be heard at diminishing intensity for a very short period of time after the source itself has stopped. This subtle prolongation of the sound in the room caused by continued multiple reflections is called *reverberation*. Because of the subtlety of reverberation effects, high background noise areas and areas with excessive amounts of acoustic softening will exhibit little if any reverberation.

Spatial connotations are associated with reverberation. A common experience is the perceptual response to a hard, empty room compared with the spatial sense of the same room after it has been softened by upholstered furnishings or by groups of people. When soft, porous materials predominate, sound reflection is significantly reduced and internal sound fades very rapidly. This *short reverberation time* (rapid rate of decay or fade) produces a spatial condition that we describe as acoustically dead. When room surfaces are consistently hard and highly reflective, on the other hand, sound continues to reflect in an unimpeded fashion between these surfaces and therefore fades more slowly. The effect of this type of space (a *long reverberation time*) is described as a live space.

Spatial Absorption

Table 2-1 indicates the various subjective listening conditions that can be created in built spaces and the corresponding mean noise reduction coefficients required. Recognize that for large, general offices, too much "softness" may actually lead to better speech intelligibility from one area to another. This is not desirable, of course, when telephone and personal conversations

Table 2-1. Room sound absorption targets for preliminary material selection.

Mean Noise Reduction Coefficient for Materials in the Space (A̅)	Subjective Listening Conditions	Suitable Use
0.4 to 0.5	"Dead" or "soft" room	Theater, lecture halls, and other spaces where use of electronic sound is intended; recording studios; spaces where significant and obtrusive background or equipment levels are present
0.25 to 0.4	"Medium" to "dead" room	Medium activity spaces, such as elementary classrooms, corridors, general offices
0.25	"Medium" room	Low activity spaces, such as private offices, small stores
0.05 to 0.25	"Medium" to "live" room	Spaces where oral communication predominates as an activity, such as conference rooms
0.05	"Live" or "hard" room	Gymnasiums, large churches or cathedrals

are to be confidential. On the other hand, too many hard surfaces in general offices can lead to too much background noise, resulting in difficult telephone and personal conversations. Acoustic conditions for general offices must be well controlled and are discussed more thoroughly later in this chapter.

Perceptual Response to Background Noise

Background noise is most often discussed within the context of sonic communication. For example, it is sometimes roughly stated (as a rule of thumb) that background noise levels of 0 to 60 dBA will permit relatively easy conversation. With background noise levels above 60 dBA, conversation becomes increasingly difficult; until, at about 115 dBA, sonic communication tends to become impossible. At this upper level, distortions within the interior of the ear will tend to mask further attempts to communicate intelligibly with high intensity signals (such as shouting into the ear of the listener). Beyond those implications associated with communication, however, moderate and high intensity background noise may also influence the perceptual response to a space utilized for *noncommunicative* activities.

The Effect of Background Noise on Noncommunicative Performance

While *pure tone* (sound at a single frequency) background noises are generally somewhat disruptive in the environment, *white noise* is a more manageable influence. *White noise* refers to a broad spectrum distribution of sound energy. With energy emitted at all frequencies, a typical white noise might be that associated with air escaping from a jet or with a waterfall. When this energy is distributed somewhat uniformly over a very wide band, it produces a *hishing* sound. When the broad band noise shifts toward higher frequencies, it becomes more of a *hissing* sound.

There have been some studies of the effect of such background noise on individual performance when auditory communication is not involved in a major way. This type of background noise can be continuous or intermittent.

Continuous White Noise

Unless the intensity is sufficient to cause pain or permanent damage, continuous white noise appears to produce no measurable disturbing effects, and apparently has little effect on participation in noncommunicative activities. Individuals can apparently adapt easily to moderate intensities of middle-frequency background noise.

There are some exceptions to this general rule, however, when higher background intensities are present and the occupant is confronted with high levels of information input (for example, when a highly complex task is being performed, or one in which considerable vigilance is required). In this situation, background noise may produce a measurable reduction in performance and reaction time. When the occupant's tasks are simplified, however, an adverse effect on performance is no longer measurable in the same sonic environment.

Since moderate levels of continuous white noise can be present without creating a disruptive influence, a selected background noise can be used or applied to establish a moderate artificial threshold. This is often useful for masking objectionable, lower intensity sounds—particularly in spaces where the need to communicate aurally is not a decisive consideration. (Also see subsequent discussion of *The Influence of Supplementary Noise Screens and Masks.*)

Continuous background noise must be handled carefully, however, because higher noise levels can produce permanent impairment of hearing. In this regard, individuals can vary considerably in their susceptibility to hearing impairment. It has been observed that at an early stage of the problem, however, damage is particularly marked at approximately 4,000 Hz. As a result, this area of the spectrum offers a potential test for early identification of those who are particularly sensitive to hearing damage. When a problem is discovered, the individual involved may be transferred to a more quiet environment; or may make use of external devices, such as ear plugs or ear muffs (hearing protectors); or environmental changes can be made to reduce the noise to a more tolerable level.

In attempting to define the general conditions where impairment problems may be expected to occur, some preliminary studies have been based on the criterion that after ten years of daily exposure, an individual should not have suffered any appreciable impairment of ability to understand speech at normal voice levels. These studies produced tentative conclusions that for an eight-hour working day, a sustained level of 85 dBA is the approximate limit that can be tolerated over a long term without producing permanent damage to hearing (table 2-2). These studies, combined with economic practicalities, have resulted in an accepted limit of 90 dBA in work environments.

Table 2-2. Background noise intensities and tolerances (based on U.S. Walsh-Healey Labor Safety and Health Standards).

Permissible Limit to Prevent Permanent Impairment of Hearing in the Speech Range	Maximum Daily Exposure (sustained period)
90 dBA	8 hours
95 dBA	4 hours
105 dBA	1 hour
115 dBA	¼ hour or less
140 dBA	maximum momentary impulse level

Example Noise Source Levels

80-dBA range: automatic appliances, such as washing machines, dryers, and dishwashers; interior of a typical residence with heavy auto traffic within 50 feet

90-dBA range: typical industrial area; typical central plant mechanical equipment at a distance of 3 to 10 feet; large cooling towers at a distance of 20 to 30 feet

100-dBA range: street with heavy truck traffic; noisy industrial processing area; typical mechanical equipment room with multiple units of central plant equipment

Intermittent White Noise

Abrupt or significant changes in background noise conditions (both increasing and decreasing intensities) can produce a startled response or reflex action, such as muscle contraction or blinking. Essentially this is explained by the fact that the listener's attention is shifted suddenly to a new information source. As the spectral band varies from the white noise and becomes narrower and more concentrated in frequency, the intermittent background noise begins to take on some aspects of pitch. As this occurs the occupant's reflex response will tend to become more intense and pronounced.

While these environmental variations have the potential negative effect of distracting occupant attention from the current focus, the listener normally has the ability quickly to process considerable information in an attempt to identify the new sound source. This characteristic has obvious intrinsic value as a warning device.

Moderate variations in a white noise background can be effective in relieving boredom. While the momentary ability to derive very high quantities of information tends to diminish after the initial reflex action, the occupant tends to be more alert and responsive in the sense that the rate of information assimilation will generally level off at a higher plateau than that which immediately preceded the response.

Annoyance

In the intensity range that lies above the lower physical limit of threshold audibility but below the limit of threshold pain, it appears that an individual's psychological tolerance for background noise in noncommunicative situations will depend on the individual's conditioning, on the individual's ability to maintain concentration at a given point in time, and on the information conveyed by the sound. In the latter category, there is evidence that noises that mystify the occupant are more likely to become annoying than sounds that can easily be located and identified by the listener. Intermittent or irregular sounds are more likely to be distracting than steady or continuous sounds. High frequency pure tones are more likely to be distracting than lower frequency tones or broad-band white noises. Furthermore, noises that seem to be avoidable, unnecessary, or inappropriate for the activity are generally found to be particularly distracting.

PERCEPTION OF SONIC SIGNALS

Sound, like light, is a medium for communication of information and ideas. Perception of meaningful sonic signals will of course depend on the emission characteristics of the source and the proximity of the source to the listener. For a given source *spectrum,* however, the listener's threshold may be a *natural* one—that is, the ear may be insensitive to some lower intensity sounds. It may also be an *artificial* threshold: the background noise level may obscure or screen lower intensity signals.

When either of these subthreshold conditions occurs, the listener will fail to perceive the affected signals. To that extent, the listener will suffer a loss of sonic comprehension, and a breakdown in communication may occur. Ideally, then, the sound contrast associated with communication activities must be sufficient to facilitate full perception and differentiation of sonic detail.

Voice Communication

Voice communication (speech) involves a variety of frequencies and intensities. Each letter or syllable is enunciated as a characteristic sound, and comprehension depends on the listener's ability to perceive each signal and distinguish it from other similar but subtly different signals.

When only 90 percent of the words are heard correctly, voice communication (listening) can become very fatiguing; and below this point, sonic communication becomes nearly impossible if accuracy or speed is required. An individual who is familiar with the language should be able to comprehend about 97 percent of the words for easy listening.

Speech Signals

Although the majority of voice signals fall within the frequency range of 250 to 4,000 Hz., optimum perception of English speech involves a total *speech range* of approximately 200 to 6,000 Hz. Voice signals emitted in normal speech also vary significantly in intensity—exhibiting a variation of approximately 25 to 35 dBA from the faintest to the more intense sounds.

Vowels are generally lower frequency signals of relatively high intensity and long duration. Consonants are higher frequency signals of lower intensity and shorter duration. The consonants generally have the greatest effect on communication because they contribute most of the actual *information* required for comprehension. Unfortunately, because they are lower intensity signals, consonants are also most susceptible to masking by background noise. As discussed earlier, the age-related reduced sensitivity to higher frequencies can contribute toward a lack of consonant intelligibility and subsequently a lack of speech intelligibility.

Intensity Variations Due to Voice Level and Distance

Figure 2-4 defines the approximate range of sounds that are emitted in speech. The intensity limits indicated in the diagram, however, will vary with the voice level of the speaker. As shown, the *conversational speech range* approximates the signals perceived by a listener standing about three feet from an individual who is speaking in a normal voice. When necessary, an

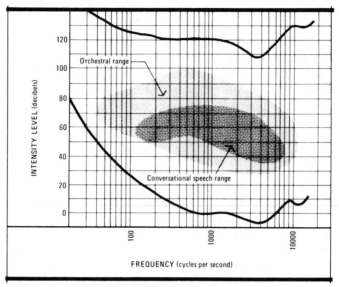

Figure 2-4. The normal sonic communication range.

increase in intensity can be produced by changing the intensity of the signal itself, or by changing the source-to-listener distance.

If the speaker's voice is lowered to a more intimate level, the entire *speech range* moves down about 6 or 7 dBA (all speech sounds become fainter in approximately equal proportions). If the speaker's voice is raised, the speech range rises about 6 to 12 dBA above the standard indicated in the diagram. If the speaker shouts, the range moves up another 6 to 7 dBA.

Direct signal intensity also varies with the distance from the speaker to the listener. For each doubling of the distance (from the three foot standard previously noted), the speech range is reduced approximately 6 dBA. The converse is also true; for each halving of the distance, the range rises approximately 6 dBA.

The Sense of Sonic Territoriality

An individual's sense of territoriality subjectively expresses the spatial volume that surrounds him. It helps to define a sense of relationship with other occupants. The precise limitations of this sense of *personal space* will vary with cultural background and with the specific activity involved; but this influence is generally present to some degree and tends to define the proximity to another person that an individual will find acceptable and appropriate under a given set of circumstances. Inappropriate territorial relationships or environmental conditions that produce unnecessary infringements on personal space can become a cause of distraction or concern among more sensitive occupants. Table 2-3 represents the general nature of

Table 2-3. Sonic territoriality.

Appropriate Audible Signal	Typical Intensity Range (decibels)	Sense of Personal Space	Physical Proximity	Implies
Soft whisper	—	Very close	3 to 6 in.	Top secret communication
Audible whisper or intimate voice	44-69	Close	8 to 20 in.	Confidential communication
Normal voice	50-75	Neutral	20 to 60 in.	Personal communication
Loud voice	56-81	Public (near)	5½ to 8 ft.	Nonpersonal communication or group information
Overloud voice	62-87	Public (across room)	8 to 20 ft.	Group address
Shouting	68-93	Upper limit	Over 20 ft.	Hailing

NOTE: Lower intensity values tend to be high frequency consonants, while higher intensity values tend to be low frequency vowels.

this psychological relationship as it helps to define appropriate signal intensities for North Americans.

Multiple Signal Vectors

When sound travels from a source to the listener by a single path (as in normal conversation at close range), relevant variations in the signal are generally clear and distinct. When greater distances are involved, however, the sound may take several paths to the listener: the direct path, reflection from one or more room surfaces, and the possible action of one or more loud-speakers. Compared to light, sound moves relatively slowly through air (approximately 1,140 feet per second). If the various major paths are significantly different in length, then, hearing problems may occur because the various vectors of the signal arrive at the listening location at significantly different times.

If various speech signal vectors arrive nearly simultaneously (in 0.035 seconds or less), the stimuli will *reinforce* each other because the arrivals are closer than the time required for the source (speaker) to emit the next distinct signal. If the individual signal vector arrivals exceed the 0.035 second spacing, the resulting interference can begin to reduce intelligibility.

When the space is small and normally proportioned, interfering reflections of this latter type are generally not significant. It is in larger rooms, where there are multiple reflections from more remote room surfaces, that problems of signal muddling and echoes may occur. (Distinct echoes will occur when the listener hears a sufficiently intense reflected sound 0.06 seconds or more after hearing the direct sound.) In these cases, for example, the first syllable of a word reaches the receiver (listener) at two distinct times—first as a syllable directly from the source and second as a syllable reflected from room surfaces. This time delay acts to superimpose one syllable over another, resulting in muddled speech and loss of intelligibility for speech perception.

The Sense of Spatial Intimacy

For communication activities that take place in moderate to large rooms (such as those intended for presentation of music or lectures), the sense of spatial *intimacy* will be affected by the interval between the time the direct sound reaches the listener and the time that the first reflected sound arrives. There are indications that, particularly for music, the optimum condition occurs when the time delay is 0.020 seconds. The sense of intimacy declines as the interval exceeds this optimum.

Reverberation Conditions for Speech and Music

Reverberating signals (reflected sounds) tend to mix with other direct and reflected sounds in an enclosed space, and this may affect the blending or clarity of various tones. So the problem of sound persistence is a basic spatial consideration in the development of an appropriate background for sonic communication—particularly in larger spaces. The objective is to provide conditions for full quality sound that is compatible with the dominant activity without excessive hardness or muddling.

In *live* spaces where reverberation times are longer than 2.0 seconds, the intelligibility of speech becomes increasingly more difficult because of a tendency for signals to become muddled and confused by superimposed reflections. Clarity of speech steadily improves below 2.0 seconds; approaching an optimum condition at about 1.0 second. No further improvement is observed below 1.0 second. There may, however, be some improvement in signal intelligibility with lower reverberation times because of the concurrent lowering of potential interfering background noise levels. Table 2-4 indicates listening conditions for various reverberation times.

Hearing conditions for music are more a matter of tradition and taste than intelligibility. For this reason, there tends to be a wide range of acceptable reverberation times. In general, however, the required ranges are summarized in table 2-5. These intervals generally apply to 500 and 1,000 Hz. Longer intervals can be acceptable at lower frequencies (and are generally desirable for most music). Figure 2-5 indicates the nature of this variation.

Background Spatial Context

Background noise in a space constitutes a context or reference level against which significant sonic signals must be compared for contrast and clarity. As a practical observation, it is generally the background noise and

Table 2-4. Reverberation time (at 500–1000 Hertz) and listening conditions.

Reverberation Time (seconds)	Listening Conditions
Below 1.0	Optimum for speech; generally too "dead" for music
1.0–1.5	Good for speech; fair for music
1.5–2.0	Fair for speech; good for music
Above 2.0	Poor for speech; fair to poor for most music; good for liturgical and symphonic music

Table 2-5. Typical reverberation time criteria (at 500–1000 Hz) for music.

Reverberation Time (seconds)	Music Situation
0.8–1.0	Relatively small rehearsal rooms
1.0–1.5	Chamber music
1.5–2.0	Orchestral music, choral music, contemporary church music
Above 2.0	Large organ, liturgical choir
Approaching 6.0	Large cathedrals

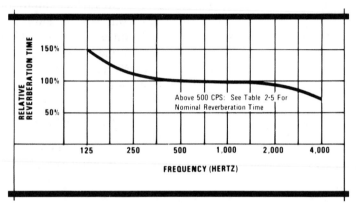

Figure 2-5. Typical reverberation time criteria for music with variations at lower and higher frequencies. (See table 2-5 for nominal reverberation time at 500–1000 cps.)

not the true physiological threshold of hearing that determines whether an individual will detect a meaningful sound. If the signal-to-noise (background) ratio is low, the marginal lower intensity sounds (such as consonants in speech) will tend to become unintelligible. Background noise at intensity levels that approximate or exceed minimum speech intensities will partially or decisively prevent comprehension of speech signals. When the signal intensity is known, the designer can expect a signal that exceeds the background intensity by 3 dB to be barely perceptible; a difference of 7 dB or more, however, is a clearly perceptible difference. In order for verbal communication to continue under adverse background conditions, it is necessary for the individuals involved to increase the intensity of their speech by moving closer together and thereby reducing the source-to-listener distance, and/or by raising voice levels in an attempt to improve sonic contrast between the signal and the background. (Fig. 2-6 shows rule-of-thumb relationships.)

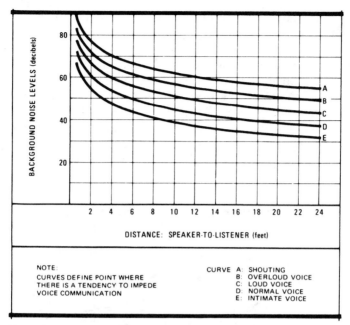

Figure 2-6. Background noise (speech interference levels).

Limitation of Reflected Noise

To some extent, internal background noise can be reduced by the addition of sound-absorbing materials and finishes. The most successful and desirable method of background noise reduction, however, is to isolate the noise source. Usually a combination of source isolation techniques and space sound absorption is used. Table 2-1 indicates the mean noise reduction coefficients of materials required in a space for a variety of environmental uses.

In general, spaces that contain operating machines or group interaction should be treated with absorbing finishes. In this way, the adverse influence of *reflected noise* is minimized; although *direct noise* in the vicinity of the source is unaffected by this procedure. Activities that produce low noise levels require less absorption. These spaces may actually utilize hard, reflective surfaces to facilitate effective distribution of reflected speech signals over or around a group of people. This need for hard, reflective surfaces becomes particularly critical if speech must be projected over distances greater than 25 to 30 feet.

Limitation of Direct Noise

When evaluating direct noise in a space, two contributing factors can be manipulated by the designer: the noise level of the equipment that must function within the space itself, and the external sound transmission through walls, ceilings, and floors. In this regard, optimum criteria are somewhat difficult to formulate as a precise numerical expression. This is because of the previously discussed need to express sonic criteria as a spectrum-type relationship between sound intensity and frequency.

When this is done, it is noted that in general the human ear can tolerate higher levels of low frequency noise, while high frequency noise must be held to lower intensities. This tolerance is true both from the standpoint of annoyance and ability to understand speech signals. *Noise and Vibration Control* (Beranek 1971) expresses this tolerance as a series of spectral background curves (fig. 2-7). These curves represent a performance standard

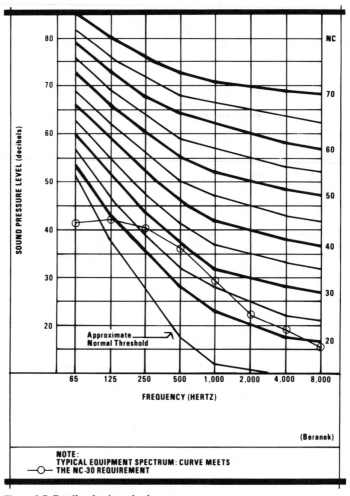

Figure 2-7. Family of noise criteria curves.

for identifying permissible background sound levels for each of the eight major octave bands. The objective is to provide a means through which a general noise level for unoccupied (but otherwise normally operating) rooms can be specified and measured in a manner that reflects both intensity and frequency (tables 2-6 and 2-7).

These background noise limitations should influence the selection of all *noise-generating equipment* (machines, air conditioning components, fluorescent and HID ballasts, and so on) and all *noise-abating elements* (such as sound attenuators and partitions). For any of these noise influences, an intensity spectrum can be developed and related to the Noise Criteria Curves. This makes it possible to estimate whether or not the source in question can be expected to become an adverse environmental influence within a given space-activity context. It is intended that for a specified criteria level, no frequency on the applicable curve should be exceeded.

Performance of Isolating Barriers and Enclosures

Sound can be transmitted from an adjacent or remote *source space* in two ways: *structural transmission,* involving actual flexural vibration of the intervening partitions; and *airborne transmission,* the direct passage of sound through openings and pores. These are essentially parallel paths, and the intensity of the transmission will depend on which of the two paths offers the least resistance.

The difference in intensity between the emitted sound in the *source space* and the perceived sound in the *receiving space* is measured in decibels and is termed *noise reduction.* For example, a 30-dB reduction is a transmission coefficient of 0.001 (that is, 1/1000 of the source intensity passes through the barrier to the receiving space); a 40-dB reduction is a transmission coefficient of 0.0001 (see table 6-1). As previously noted, however, any meaningful analysis of sonic factors must evaluate the full spectrum of intensities and frequencies.

Sound Transmission

Sound Transmission Class (STC) is a useful device for preliminary evaluation and prediction of noise transmission through a barrier. This method relates to common sounds, such as those emitted in human speech, in music, by a barking dog, and so on. Airborne energy from these signals will induce subtle vibrations in a wall, ceiling, or floor. This action in turn induces sound in the adjacent space. STC, then, is a method for specifying the performance of a barrier in resisting the transmission of these noises. (Also see discussion of sound transmission in chapter 6.)

Prespecified spectral contours are defined in figure 2-8. For a given barrier, the measured transmission loss (TL) values at each of the frequency

Table 2-6. Background noise (typical criteria).

Background Noise Levels (see fig. 2-7)	Sonic Conditions	Suitable Use
Below NC-25	Very quiet	Restful, contemplative
NC-25-35	Quiet Normal conversation: 10-30 ft	Reception, discussion, classroom, office, general home
NC-35-45	Moderately noisy Normal conversation: 6-12 ft Raised voice: 10-30 ft	General office (limited discussion)
NC-45-50	Noisy Normal conversation: 3-6 ft Raised voice: 6-12 ft Telephone use becomes slightly difficult	General office (w/o machines), drafting
NC-50-55	Noisy Normal conversation: 1-2 ft Raised voice: 3-6 ft Telephone use becomes difficult	Typing, clerical (w/machines)
NC-55-70	Very noisy Raised voice: 1-2 ft Telephone use becomes unintelligible at higher levels	Industrial

centers shown in figure 2-8 are plotted. The STC rating is then determined in a trial and error process according to the following method:

1. No plotted-test transmission loss curve point can be more than 8 below a selected STC rating curve.

2. The sum of the deficiencies at all points of the plotted-test transmission loss curve cannot be greater than 32.

The example in figure 2-8 illustrates a wall construction having a STC rating of 34. Table 2-8 reports various STC ratings and subsequent sonic effects.

Impact Noise

Impact noise is sound generated by an object striking, vibrating, or sliding against a component of the building. This includes the effect of footsteps

Table 2-7. Noise criteria targets (background noise limits).

dBA	Noise Criteria Curve (see fig. 2-7)	Sonic Environment
<30	NC-15-20	Concert halls, broadcast studios
30-34	NC-20-25	Assembly areas without amplification, theaters, auditoriums, courtrooms
30-38	NC-20-30	Sleeping spaces, such as hospital rooms, hotel and motel rooms, apartments
34-42	NC-25-35	Classrooms, music rooms
38-42	NC-30-35	Churches, library reading rooms, private offices, conference rooms
42-47	NC-35-40	Heavy occupancy areas where limited discussion occurs, such as stores, gymnasiums, cafeterias, large offices
47-52	NC-40-45	Heavy circulation areas, such as lobbies, corridors, large offices
52-56	NC-45-50	Machinery and equipment spaces, such as garages, laundries, equipment rooms
56	NC-50	Assembly areas with amplification

(foot-fall), moving furniture, dropped objects, doors slamming, mechanical equipment vibration, and so on. Such actions will set the affected barrier or enclosure in vibration, which induces an impact sound on both sides of the barrier.

Impact Insulation Class (IIC)

Impact Insulation Class is a useful system for quick evaluation and prediction of the ability of a given floor-ceiling construction to resist and subsequently reduce common impact sounds. The system is devised to be comparable to the air-borne sound STC ratings. A scale is created wherein a higher rating number represents a greater capability by the tested floor-ceiling system to resist structurally borne or impact sound transmission. Figure 2-9 defines the standard IIC contour.

Similar to the STC contour, the IIC contour is moved vertically with respect to the measured data. Again,

1. A single unfavorable deviation may not exceed 8 dB.
2. The sum of the unfavorable deviations may not exceed 32 dB.

Table 2-8. Typical partition sound transmission criteria and noise criteria combinations.

Partition Transmission Loss (STC)			Sonic Conditions in the Receiving Room	Typical Applications
NC 25[1]	NC 35[1]	NC 45[1]		
STC 35 (or less)	STC 30 (or less)	—	Normal speech can be easily and distinctly understood through the wall	Privacy not required; partition used only as a space divider
STC 35-40	STC 30-35	—	Loud speech (and loud radios) understood fairly well through the wall; normal speech can be heard, but not easily understood	Suitable for dividing noncritical areas; provides fair degree of freedom from distraction
STC 40-45	STC 35-40	STC 30-35	Loud speech can be heard through the wall, but is not easily intelligible; normal speech can be heard only faintly, if at all	Provides good degree of freedom from distraction; suitable for junior executives, engineers, apartment dwellers, and so on
STC 45-50	STC 40-45	STC 35-40	Loud speech can be heard faintly through the wall, but is not easily understood; normal speech is inaudible through the wall	Provides a confidential degree of conversational privacy; generally suitable for doctors, lawyers, senior executives, and apartment dwellers
STC 50	STC 45-50	STC 40-45	Loud sounds such as brass musical instruments and singing, or a radio at high volume can be heard only faintly through the wall	Suitable for dividing private or relaxing spaces from noisy adjacent spaces that house loud activities or equipment

1. NC values describe the background noise levels in the receiving room (space of interest). Refer to discussion of Noise Criteria.

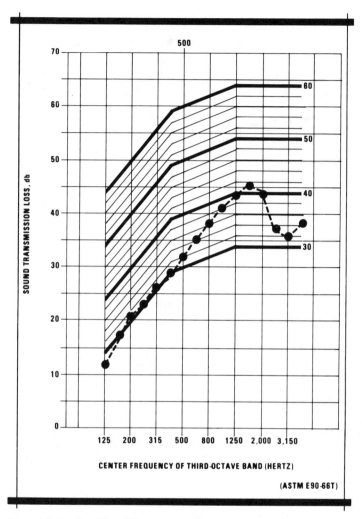

Figure 2-8. Family of Sound Transmission Class (STC) curves.

IIC curves are generated for available floor-ceiling assemblies by using a "tapping machine" to generate impact sounds on the actual assembly. Sound pressure level measurements are then taken in the receiving space, and the results are plotted according to the vertical left scale. In order to have a standard number rating, the actual IIC value is read from the right-hand scale. This scale subtracts the measured sound level from 110 dB to give the IIC rating.

Readings are determined at one-third octave bands with center frequencies in the range from 100–3150 cps. The IIC rating is read on the IIC scale corresponding to standard contour position at 500 cps.

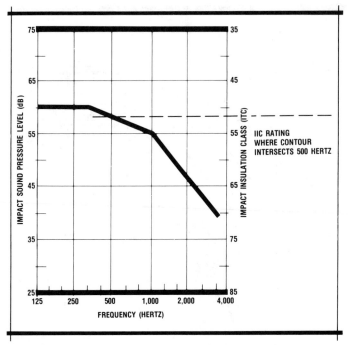

Figure 2-9. Impact Insulation Class (IIC) contour.

Table 2-9 lists a variety of floor-ceiling situations and corresponding IIC target values. Recognize that the noise generated in the receiving space affects the IIC value required of the floor-ceiling assembly. Building-material manufacturers generally have available a variety of floor-ceiling assemblies and their corresponding IIC values to help the designer meet the IIC targets required for various projects.

The Influence of Supplementary Sound Screens and Masks

While previous discussions have dealt primarily with the negative aspects of background noise in reducing the sonic contrast of signals intended for communication, background noise also affects the psychological sense of *privacy*. In this respect, the audible background level is referred to as background sound—sound connotating a helpful sound source, while noise connotates a distracting sound source. For example, an office in a rural or residential area may be classified as *noisy* by people who are accustomed to the higher background levels of sound commonly found in downtown locations with background noise stemming from mechanical systems in or on nearby buildings and heavy street traffic. The description *noisy* refers, in this instance, to distraction or lack of sonic privacy—the sound generated

Table 2-9. Criteria for IIC for floor–ceiling assemblies between dwelling units within average noise-level environments.

Lower Room (noise receiver)	Upper Room (sound source)				
	Bed-room	Living Room	Kitchen	Family Room	Corridor
Bedroom	52	57	62	62	62
Living Room	52	52	57	60	57
Kitchen	50	52	52	58	52
Family Room	48	50	52		
Corridor					48

SOURCE: HUD: *Airborne, Impact, and Structure Borne Noise — Control in Multifamily Dwellings.*

by mechanical systems and street traffic in downtown locations actually helped to *mask* office noises (such as copy machines, typing, office chatter, and telephone ringing). In a relatively quiet space, low intensity sounds and irrelevant distant conversations may be easily heard; this becomes distracting.

Background or *ambient* or *masking* sound, then, can serve a positive function by establishing an artificial threshold that will mask remote or low-intensity distractions. Such a noise mask will support a sense of sonic privacy in a large open space, and it can also supplement the attenuation value of partitions.

Figure 2-10 suggests how a partition that is inherently only moderately effective as a barrier to sound transmission may be completely suitable when a continuous background sound becomes an innocuous part of the sonic environment. But when this ambient sound screen is reduced or removed, the artificial threshold lowers and the same partition may be inadequate for effective sonic isolation. Care must be taken in developing the background masking sound system. Generally the only successful method involves an electronic sound generating system that is specifically tuned to the acoustical characteristics of the space.

Intermittent intensities and pure tones are generally not appropriate as an effective mask or screen because they may in themselves introduce an adverse influence into the environment. High frequencies are particularly troublesome in this regard. If either one of these must be present in an environment intended for sonic communication, however, it is generally desirable to reduce the source intensity so that the critical frequency falls 5 to 10 dB below the applicable Noise Criteria Curve (see table 2-7). If the noise is both intermittent and pure tone, the acceptable background level should fall 10 to 20 dB below the usually applicable curve.

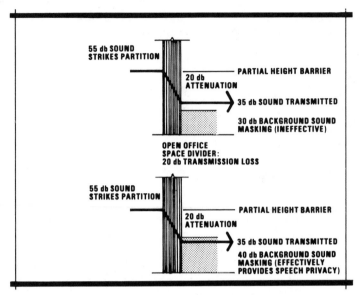

Figure 2-10. Typical effect of background noise screens.

TOTAL SYSTEM PERFORMANCE:
THE OPEN OFFICE SONIC ENVIRONMENT

With the growing popularity of the open office plan concept in the late 1960s and early 1970s, a new metric was developed in an attempt to describe the level of total acoustical system performance. Factors affecting the sonic environment in an open office include sound transmission and supplementary background sound masking (see previous discussions in this chapter). Sound transmission in the open office depends on the ceiling system (its "softness" and the type of luminaires present, if any), the space dividers (their height, "softness," transmission capability, and layout within the office), and occupant-to-occupant distance. Sound masking depends on the selection of background masking system components (the electronic sound generating equipment and speaker placement and quantity). Recognize that background sound masking becomes so important in an open office environment that the designer cannot reliably depend on the HVAC system sound to fulfill this background sound requirement.

Through its ambitious building systems approach to design and construction in the early 1970s, the U.S. Government's Public Buildings Service (PBS) sponsored the development of acoustical performance criteria for open office systems (PBS 1972). A single number criterion, designated "SPP" for Speech Privacy Potential, was introduced to evaluate the combined effects of three components on open office acoustics—the space dividers, ceiling system, and the background sound masking system. A complete "system" of these three components with an SPP rating of 60 or more is considered a good open office system capable of creating an environment where speech intelligibility from one workstation to the next will be difficult if not impossible. This recognizes the fact, however, that voices may be heard from workstation to workstation, but that word and sentence recognition will be unlikely. Hence, in a large, open plan office with a 10'-0" ceiling height or less, SPP 60 is achieved when 1½" fiberglass ceiling pads are used, *and* deep/wide-cell (3" by 3" or greater) parabolic or narrow-width indirect luminaires are used, *and* 5'-5" or higher acoustical space dividers (partitions) are used with a core of aluminum foil sandwiched between two layers of 1½" fiberglass board are used, *and* a background masking sound system is used.

Although this particular criterion and the measurement and calculation procedures behind it have been disputed over the past decade, no new significantly different criteria and/or procedures have yet surfaced that are as widely used.

REFERENCES

Beranek, Leo L. 1971. *Noise and Vibration Control.* New York: McGraw-Hill.

Berendt, Raymond D., Winzer, George E., and Burroughs, Courtney B., September, 1967. *Airborne, Impact, and Structure Borne Noise—Control in Multifamily Dwellings,* U.S. Department of Housing and Urban Development, Washington, D.C.

Peterson, Arnold P. B., and Ervin E. Gross. 1974. *Handbook of noise measurement,* 18-19. Concord, Mass.: General Radio.

Public Buildings Service. 1972. *Test method for the direct measurement of Speech Privacy Potential (SPP) based on subjective judgment,* PBS-C.1. Washington, D.C.: U.S. Government, General Services Administration, Public Buildings Service, Office of Construction Management.

Yerges, Lyle F. 1969. *Sound noise and vibration control,* 10-11. New York: Van Nostrand Reinhold.

CHAPTER 3

The Thermal Environment

THERMAL REACTIONS OF THE HUMAN BODY

Factors involved in the thermal environment are not directly related to communication or orientation. Unlike some of the more subtle considerations related to the perception of visual or sonic detail, thermal factors exert a relatively minor influence on individual participation and performance — *unless* these factors become adverse to the extent that they induce physiological stress.

The human body receives chemical energy from food. This energy is, in turn, transformed into other forms of energy by the body; and heat is released by this *metabolic process.*

Metabolic rate (that is, the rate of heat emission) will vary significantly with movement or activity and type of clothing. To a less significant degree, it will vary with the size, age, and sex of the individual.

For a given person who is steadily involved in the same activity, however, it appears that his or her metabolic rate will be stable in the environmental temperature range of approximately 60°-90°F. Below this range, the adaptive effect of shivering may begin to increase the metabolic range. Above the stable range, the metabolic rate may again tend to increase, not as an adaptive procedure, but because the internal chemical action will increase as the deep body temperature moves higher.

The heat flow interaction between the body and the surrounding environment takes place through:

- Convection
- Conduction
- Radiation
- Evaporation

The direction of heat flow may be either to or from the body. If the algebraic sum of heat flow quantities is either a net flow to the body, or a net flow from the body at a rate below the metabolic rate, this implies a heat storage condition within the body that will tend to produce an increase in the temperature of the deep body tissues. Similarly, a net flow from the body at a rate that exceeds the metabolic rate will produce a decrease in the deep body temperatures. If either of these actions is continued over even a moderate period of time, adverse physiological effects (including death) will result.

Methods of Body Heat Adjustment

When in a relaxed state, a human occupant will generate approximately 400 Btu/hr. Figure 3-1 indicates the general method by which this body heat is removed and dissipated.

The dissipation curve begins to drop sharply as the environmental temperature approaches and exceeds the deep body temperature of 98.6°F. This indicates the diminishing capability of the body to cool itself effectively in these environmental circumstances. The result of this condition is eventual heat exhaustion.

In the range of environmental temperatures below approximately 70°F, additional body heat is indicated in the diagram. This reflects the adaptive effect of involuntary physical activity (such as shivering). As this becomes more pronounced and excessively prolonged, the individual can begin to suffer from exhaustion and overcooling of the skin (exposure).

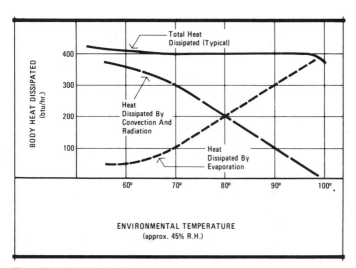

Figure 3-1. Body heat exchange.

Adjustments to Warmer Environments

When the environmental air temperature lies within the neutral zone for relaxed human occupancy (typically the low to mid-70° range), the flow of blood through the body utilizes the deep-seated arteries and returns through the deep veins. As a result, relatively little heat is transferred from the blood to the surface skin.

As the environmental temperature begins to rise above the neutral zone, the superficial veins (particularly those in the limbs and extremities) begin to dilate. The diversion of returning blood to these superficial veins causes the blood to act as a heat-convecting medium, increasing the flow of heat from the deeper tissue to the surface and causing the skin temperature to rise. This action increase the loss of heat at the skin by radiation and convection to the external environment.

When dilation of the veins reaches the normal maximum, skin temperature is somewhat uniform over the entire body and has reached a normal maximum. The body is, therefore, incapable of increasing body heat dissipation by convection and radiation. If environmental temperatures continue to rise, the skin pores tend to open and sensible perspiration becomes a mechanism to increase body heat dissipation through evaporative cooling of the skin.

The rate of perspiration will vary with the heat dissipation requirements. At approximately 98.6°F, where convection and radiant losses are negligible because there is no essential temperature difference between the human circulation system and the surrounding environment, evaporation tends to become the exclusive body regulative process.

The perspiration mechanism tends to fatigue with time. So the rate of perspiration will decrease after several hours of exposure to adverse conditions. This fatigue is partially related to the fact that the perspiration process consumes body fluids. (Approximately one pound of body fluid is consumed for each 1,000 Btu dissipated, resulting in losses in body weight.)

Adjustments to Cooler Environments

When environmental temperatures drop, the previously discussed processes are, of course, reversed. When temperature falls into the neutral zone, then, the pores tend to close, so relatively little heat is dissipated through evaporation. Most of the body heat is being dissipated by convection and radiation. Experiments indicate that radiation losses constitute the most important single means of dissipation in this environmental range.

As temperatures fall below the neutral zone, further physiological changes occur as the body attempts to minimize heat losses due to radiation and convection from the skin surface. The superficial veins become constricted, while the deeper veins dilate. Body heat is therefore retained within the deeper tissues, while the skin surfaces (particularly the extremities) are permitted to cool. When this drop in extremity surface temperature is significant, the loss of heat by radiation and convection tends to decrease.

As environmental temperatures continue to drop and the variation in blood flow can no longer compensate adequately, deep tissue temperatures drop and involuntary physical activity (shivering) becomes a means to generate additional body heat. This action does not fully compensate for the drop in deep tissue temperature, but rather attempts to prevent a continued drop.

Acclimatization

As environmental temperatures rise or fall and stabilize at new levels (as may occur with seasonal changes), the more permanent constriction or dilation of the superficial veins requires the body to respond by varying the total volume of blood. This variation may amount to as much as 20 percent, with the greater volume required for the warmer environmental conditions.

These adjustments in blood supply will require approximately four days or more to accomplish, and this is a major factor in determining the general *acclimatization* of the body to warm or cool climatic conditions. Acclimatization will also affect sweating ability. With continued exposure to high temperatures, the maximum sweating rate will tend to increase over a period of several weeks.

Variations in Activity Level

The previous discussions have described the action of a typical human occupant in a relaxed state. In this condition, body heat is generated at approximately 400 Btu/hr.

Table 3-1 shows how metabolic rate increases with increasing activity. When the occupant is participating in more strenuous activities, then the rate of body heat dissipation is increased dramatically.

This extreme variable rate of body heat generation suggests that identification of a *neutral zone* in the environment varies with the nature of the activity being performed. An environmental temperature of 65°F may be excessively warm for an active group of men doing heavy manual work; but the same condition may be quite cool for a group involved in more leisurely activity.

Table 3-1 also includes an attempt to approximate neutral comfort ranges for the various activity classifications. In each case, the temperatures shown assume that the participating individuals will remain involved for a moderate period of time.

Table 3-1. Typical effects of activity variations (thermal).

Activity	Approximate Body Heat Generation (Btu/hour)	Typical Neutral Zone[1] (environmental air temperature)
Heavy exercise	1,500-2,500	55°-60°
Moderate manual work	750-1,500	60°-65°
Normal circulation	600-750	65°-70°
Normal rest—seated[2]	400	70°-75°
Normal rest—bed	250	75°

1. Applies for typical individuals who will tend to remain under these activity conditions for at least 2-3 hours.
2. See figure 3-1.

Skin Temperatures

Skin temperatures vary considerably. In the normal range of environmental temperatures, the average skin temperature of the occupant at rest may vary from a maximum of approximately 97°F to a minimum of 86°F without any change in deep tissue temperature. The skin temperature of the extremities may drop even lower under normal environmental conditions.

As the environment warms above optimum, then, an early effect is that both trunk and extremity skin temperatures rise toward the maximum values. (This action increases the dissipation of heat through convection and radiation.) Further warming causes sensible perspiration.

Optimum comfort for clothed individuals at rest appears to occur when the average skin temperature is in the range of 91-93°F. For active individuals, this optimum value is lower, sometimes falling as low as 85°F when heavy exercise is involved.

The Sensation of Temperature Change

In addition to effects associated with metabolic processes, the body also reacts to external hot or cold stimuli of a spatial or localized nature. This sensation of warmth or coolness depends on changes in skin temperature.

However, the body is not a good *thermometer* for estimating the intensity of temperature variations. The general stabilizing effects of acclimatization may affect such judgments. Furthermore, the skin is more sensitive to heating than cooling. Some research groups have reported that a warming sensation is produced by a rate of increase in skin temperature of 0.002°F/sec or more; while a cooling sensation requires a decrease in skin temperature of at least 0.007°F/sec.

Response to Humidity

Dissipation of body heat though evaporation is directly proportional to the quantity of perspiration that is evaporated. Since the evaporation rate will vary significantly with the saturation of the surrounding air, relative humidity is a particularly critical environmental factor that affects the ability of the body to dissipate heat when environmental temperatures are high. As a corollary, relative humidity will affect the highest range of environmental temperatures to which the body can be exposed (in still air) and still maintain effective heat regulation through evaporation. One study has reported typical limits as follows: 88° @ 100% RH, 99.5° @ 51% RH, 113° @ 18% RH, and 126° @ 0% RH.

As environmental temperatures cool, however, the influence of humidity on the ability of the body to dissipate heat becomes less significant. When the pores are closed and sweating action is no longer involved, humidity only affects heat that is dissipated through breathing. This heat transfer depends on the taking in of relatively unsaturated air, with subsequent expelling of saturated air. The greater the difference in saturation, the greater the body heat loss; so this heat transfer is greatest when environmental humidity is low. In most cases, however, heat dissipation through this means is less significant than through other methods previously discussed.

The Total Thermal Environment

In summary, the human body functions as a heating and cooling system. It can, within limits, effectively and instinctively adjust to environmental conditions that vary from optimum. Essentially, the body loses heat at a controlled rate to moderately cooler air and cooler surfaces. If this cooling action is either too slight or excessive, however, the result can be occupant discomfort and physiological stress.

The thermal environment is much more complex, though, than simple temperature relationships. The occupant also emits heat through exhalation and through evaporation of perspiration. As a result, the body responds to the quality of air in the space, both in terms of *freshness* and humidity. Ventilation, filtering, and water vapor control are thus additional criteria that relate to the heat transfer potential of convection and radiation.

Thermal comfort is a near-optimum condition of equilibrium between the human body and an environmental background. The environmental background must facilitate favorable atmospheric interchanges, maintain a condition for a favorable skin temperature, and minimize the need for any of the more extreme physiological adjustments that are required to maintain a balance between internal body heat gains and external heat losses. In this way, occupants of built space will be freed of thermal and atmospheric stress, and, assuming that the luminous and sonic environments are equally satisfactory, be able to perform in an optimum manner.

PERCEPTION OF THE THERMAL-ATMOSPHERIC BACKGROUND

When the designer defines the performance characteristics of the thermal environment, he or she generally strives to achieve and maintain a somewhat neutral background condition. The initial objective is to minimize adverse thermal and atmospheric influences that could otherwise impede either participation in an activity or effective performance of a task.

In the sense of system management, this requires manipulation of several related environmental factors:

- Temperature and humidity of the air mass that surrounds the body
- Movement and composition of the air mass
- Temperature of the major surfaces that surround the body

The appropriate manipulation of these factors with regard to the activities to be performed and the culturally conditioned clothing standards in a given space should result in thermal equilibrium for the occupants.

Environmental Air Temperature and Humidity

The relationship of the thermal variables of environmental air temperature and humidity are graphically represented in the psychrometric chart, figure 3-2. The horizontal scale defines environmental dry bulb temperature, that is, the temperature of the environmental ambient air. The vertical scale defines vapor pressure or the quantity of moisture present in a given air mass. The curved lines represent relative humidity (RH) or the percentage of moisture in a given combination of dry bulb temperature and moisture quantity relative to what the air mass could hold at saturation (100% RH). Effective temperature, ET*, refers to any set of temperature and humidity conditions that gives the same sensation of comfort as the stated temperature at 50% RH. The effective temperature relationship applies for still air conditions (15-25 feet per minute), when the occupants are seated at rest or doing very light work, wearing typical American indoor clothing, and when the room surfaces are at or near the environmental air temperature. Examples of effective temperature can be found in the next section where the boundaries of the thermal comfort zone are defined.

The Comfort Zone

Figure 3-2 identifies a *thermal comfort zone,* or conditions at which at least 80 percent of the occupants of a space would find the thermal environ-

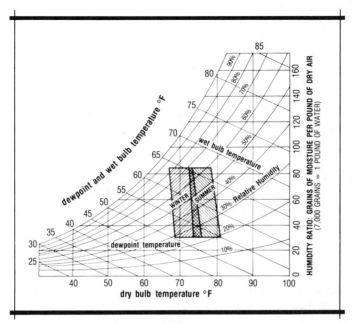

Figure 3-2. Psychrometric chart indicating winter and summer comfort zones.

ment acceptable. Again, the occupants must be doing light work or at rest and dressed in typical American indoor clothing, the air motion must not be greater than 30 feet per minute in winter and 50 feet per minute in summer, and the room surfaces must be at or near the environmental air temperature. The vertical boundaries of the winter zone are at 68°F and 74.5°F ET* and of the summer zone 73°F and 79°F ET*.

The top and bottom of the comfort zone are defined by dewpoint temperatures (DPT). DPT is defined as the temperature at which moist air becomes saturated (100% RH) with water vapor when cooled at a constant pressure or water content. The top of the comfort zone is at 62°F DPT and the bottom of the comfort zone is at 35°F DPT. These limits are defined by limits of health, mold formation, and human comfort. The impact of humidity level upon sedentary human comfort is not as severe as the impact of temperature.

This standard can and should be adjusted for different conditions of activity, clothing, air movement, and surface temperatures as well as localized nonuniform conditions. Isolated environmental conditions frequently occur which do not conform to the overall designed conditions due to inadequate air distribution, lighting "hot" spots, or human exposure to hot or cold radiant surfaces such as a window. The boundaries of this comfort zone in reality are not as sharp as depicted and must be analyzed with respect to the unique conditions of each design program.

Variations from the Assumed Conditions

Variations from the general conditions defined in figure 3-2 are as follows.

Densely Occupied Spaces

For congested occupancy conditions, comfort will occur at somewhat lower temperatures. This change is due to the effect of radiant heat transfers between occupants.

Activity

Occupants who are involved in more strenuous activities will also require lower temperatures. See table 3-1.

Ages of the Occupants

Studies have revealed that the thermal environments preferred by older people do not differ from those preferred by younger people. The lower metabolism in older people is compensated for by a lower evaporative loss. In practice, the ambient temperature level in the homes of older people is often higher than that for younger people. This may be explained by the lower activity of elderly people, who are normally sedentary for a greater part of the day.

Adaptation

It is likely that the same comfort conditions can be applied throughout the world. In determining the preferred ambient temperatures, however, local clothing habits and styles should be considered.

Sex

Men and women prefer almost the same thermal environment. Women often prefer higher ambient temperatures, however, due to the fact that they often wear lighter clothing.

Local Thermal Discomfort

Thermal comfort requires that no local warm or cold discomfort exists in any part of the human body. Such local discomfort may be caused by an asymmetric radiant field such as cold windows or warm heat sources, local convective cooling or drafts, contact with warm or cold floors, or by a vertical temperature difference between the head and feet.

Geographical Variations

As previously discussed under "Acclimatization," our perception of thermal comfort will vary with the season and geographic location. This is due to physiological responses expressing themselves in subjective judgment. The designer should recognize the conditioning and varying responses of the occupants of the space.

Condensation

Condensation is a factor in any evaluation of temperature and humidity. This condition will occur when the temperature of any surface falls below the dew point of the ambient air. For a constant interior humidity and air temperature, the point within a building envelope system at which condensation will occur varies *directly* with the inducing temperature influence and *inversely* with the heat transmission characteristics of the materials. The better the insulation value, the more remote and intermittent is the problem of condensation.

For this reason, thin metal and glass (both of which conduct heat quite readily) are most susceptible to condensation. Cold air ducts, metal air supply diffusers, cold water pipes, and window sashes are particularly vulnerable to this problem, as is a single pane window glass. Table 3-2 indicates at what temperatures condensation will form on the inside face of typical building envelope systems for different interior relative humidity conditions.

When condensation is likely, the problem can be controlled in different ways:

• Reduce the relative humidity of the interior air mass. This action will lower the temperature range at which the problem will occur, but it may also

Table 3-2. Approximate outside temperature (°F) at which condensation will begin to occur on the inside face of the material, based upon the thermal insulating values of the materials.

Envelope Insulating Material	*Interior Relative Humidity (70° interior air)*			
	20%	*30%*	*40%*	*50%*
Winter condensation problems				
Aluminum or steel	16°	28°	38°	46°
Glass: single glazing	13°	25°	36°	45°
Minimal winter condensation problems				
8-inch face brick wall	—	−17°	3°	22°
glass: double glazing	—	−20°	1	20°
6-inch concrete roof	—	−26°	−3°	17°

reduce occupant comfort and well-being if the humidity is maintained too low. It is generally desirable to maintain relative humidity at a level of 20–30% or higher.

• Increase the insulation value of the envelope system. This will also lower the exterior temperature range at which the problem will occur. This is accomplished with an air space (such as in double or triple glazing systems) or by adding insulation within or on the system itself.

• Provide a vapor barrier between the warm interior ambient air and the position at which the dewpoint temperature occurs within the building envelope systems. See later discussions for examples of the analysis of the location of this vapor barrier.

Air Distribution and Composition

As discussed previously in this chapter, a human occupant in a thermally neutral environment will emit heat to the surrounding air. This action will, in turn, raise the temperature of this air. Furthermore, evaporation of perspiration and support of the breathing process will modify the composition of the air mass.

Machines such as video display terminals, typewriters, audio-visual equipment, copiers, and other heat-producing devices will add heat to the air mass. Also, the building envelope thermal transmission will affect the air mass. By-products from building materials and the air-handling system can also produce additional components which affect the overall air mass composition and quality. Although not the only means of thermal control, air mass composition and air movement are particularly significant in affecting many aspects of the thermal-atmospheric environment.

Compensating Effects of Air Motion

When air temperature and humidity exceed the human comfort range, increased air motion can become a compensating influence and return the sensory environment toward a condition of thermal equilibrium. In this sense, increased air velocity expands the thermal comfort zone.

Table 3-3 indicates how the human occupant will perceive various levels of air motion. Table 3-4 indicates how increased air velocity can expand the comfort zone by compensating for excessive temperatures and humidity.

Indoor Air Quality

Because more than 90 percent of the average person's time is spent indoors and pollutant concentrations can be many times the levels in the outdoors, indoor air quality has become a critical issue in recent years. Tested occu-

Table 3-3. Background air motion and typical behavioral patterns.

Air Speed at Head Level (feet per minute)	Subjective Evaluation
less than 15	Complaints about stagnant air when other atmospheric conditions lie in the comfort range
15-25	Favorable; basis for *comfort zone* (figure 3-2)
25-50	Favorable conditions when atmospheric conditions lie in the comfort range
50-100	Subtle awareness of air movement, but generally comfortable when temperature of the moving air is at or slightly above room air temperature
100-200	Constant awareness of air movement, but generally pleasant
200-400[1]	Increasingly drafty conditions but favorable cooling provided in warm conditions
400-700[1]	Adverse effects in disrupting a task, an activity or personal composure

1. Localized and intermittent introduction of moderate velocity room temperature air at head level will introduce temporary stimulation in an occupant who is involved in long-term activities or work tasks.

pants of indoor environments often equate indoor air quality and thermal comfort in responding to questions of thermal acceptability and comfort. It is a difficult issue with which to deal because (1) the effects of poor air quality upon humans vary much from individual to individual, (2) the impact is often not immediate, sometimes occurring hours later in different environments or, in the case of cancers, not occurring until years later, and (3) the exact causes are difficult to determine. Spatial quality, lighting, acoustics and general individual psychological conditions can confuse the responses from tested occupants.

Composition of Air

In the normal human body process, oxygen is taken in and consumed, and carbon dioxide and water vapor are expelled. An individual at rest may take in 16.2 cubic feet of air per hour (CFH)—consuming 0.504 CFH of oxygen and producing 0.42 CFH of carbon dioxide.

As the activity becomes more strenuous, the required quantities of air increase dramatically. For example, an individual walking at a rate of 4 MPH will take in about 78.6 CFH of air—consuming 3.38 CFH of oxygen and producing 3.0 CFH of carbon dioxide.

Table 3-4. Background air motion (compensation for rise in air temperature and humidity).

Environmental Conditions	Air Velocity Required to Approximate Near-Optimum Body Heat Transfer Conditions for an Occupant at Rest (feet per minute)
76° 50% RH	15-25
80° 35% RH	100
75° 80% RH	100
82° 30% RH	200
75° 100% RH	200
85° 35% RH	700
79° 100% RH	700

Table 3-5. Air composition (tolerance).

	Percent by Volume	
	Carbon Dioxide	Oxygen
Normal air condition	0.3	20.6
Objectionable condition	2.0	16.0
Dangerous condition	5.0	12.0

NOTE: 1.5 percent CO_2 is considered permissible for submarines. Approximately 3.5 cfm of outdoor ventilation air per occupant will usually maintain the composition at near normal conditions.

High occupant densities or a prolonged period of occupancy can therefore begin to alter the composition of air in a sealed space (such as in a mine, a submarine, an aircraft, or a tightly sealed building). When this occurs, the quantity of oxygen is reduced, while the carbon dioxide increases. Table 3-5 shows the relationships that define both normal and abnormal conditions.

Generally, a normal balance is preserved by adequate ventilation. As a general rule of thumb, 1.0 cubic feet per minute (cfm) of ventilation air per occupant will preserve the oxygen balance; and 3.5 cfm of ventilation air per occupant will preserve a CO_2 content below 0.6 percent. However, effective control of body and other odors and the mixing of outdoor air and indoor air for achievement of acceptable noncontaminated air requires higher quantities of ventilation air. (See table 3-6.)

Table 3-6. Indoor fresh air guidelines.

Application	Estimated People/ 100 sq. ft.	Outdoor Air Requirements	
		cfm/person	cfm/sq. ft.
Auditoriums	15	15	
Classrooms	5	15	
Dining facilities	7-10	20-30	
Enclosed parking garage			1.5
Hospital patient room	1	25	
Hotels, motels, resorts, dormitories		30 cfm/room	
Offices	0.7	20	
Public corridor			0.05
Public restroom		50 cfm/wc or urinal	
Retail, street floor	3		0.3
Theater	15	15-20	
Residential	0.35 ACH but not less than 15 cfm/person		

SOURCE: *ASHRAE STANDARD 62-1989.* "Ventilation for Acceptable Indoor Air Quality."

The ventilation rates established are based on the concept that an acceptable level of *indoor air quality* (IAQ) can be achieved through the introduction of outdoor air which will be mixed with the indoor air. This mixture then is a reasonable control of carbon dioxide, particulates, odors, and other contaminants common to the environment and the described function.

Care must be taken that the outdoor air is not in itself contaminated. If there is a concern that the outdoor air is contaminated, the air must be tested and, if necessary, appropriate purification techniques utilized. Also, the outdoor air must be drawn from points which are not vulnerable to cross-contamination by exhausted indoor air, by stagnant conditions such as standing water or dead air, and by exhaust from combustion sources. The location of the fresh air intake grille is, then, critical to the operation of the HVAC system. Care must be taken to recognize exterior air conditions and prevailing wind patterns.

Atmospheric Impurities

Control of atmospheric contaminants is an important aspect of air handling. Control affects comfort, and in some cases, the health of the occupants.

Essentially, this aspect must consider three factors of air composition: oxygen and carbon dioxide, odors, and air contaminants. They are controlled through the introduction of fresh outside air, the recirculation and treatment of the interior air, and the maintenance of adequate air flow to all portions of a space.

Indoor air quality is established by limiting contaminants to specified, acceptable levels. In addition, a subjective evaluation by the occupants will establish acceptability of the air as to odors and discomfort such as irritability of the eyes, nose, and throat. The outdoor air ventilation rates of table 3-6 yield acceptable indoor air quality unless carcinogens or other harmful contaminants exist in either the indoor or the outdoor air. If there is suspicion that such contaminants may be involved, in-depth analysis and study must be made to control such elements at recommended levels.

The problem for the designer is to identify potential contaminants, their sources, and to find ways to control and dilute the contaminants. The contaminant sources may be in the outside air, the recirculated air, or within the environment itself. Table 3-7 lists typical pollutants and their possible causes.

Energy conservation issues have promoted reduced circulation of air and limited the introduction of fresh air. Although contributing to energy conservation, this effort has also contributed to the "tight building syndrome," wherein the building contains and holds pollutants for long periods. For example, outgassing of new materials in tight buildings has contributed large amounts of pollutants which otherwise would have been eliminated. Studies by the National Institute for Occupational Safety and Health (NIOSH) have indicated that 17 percent of identified indoor air quality problems occurred due to contaminants released within the building, 11 percent due to contamination from outside the building and 52 percent due to inadequate ventila-

Table 3-7. Indoor air pollutants and possible sources.

Pollutant	*Possible Source*
Environmental tobacco smoke	
Radon	Natural earth materials
Respirable particles	Stoves, fireplaces, asbestos, fiberglass
Formaldehyde	Insulation, particle board, surface materials
NO_2, SO_2, CO	Stoves, fireplaces, water heaters
Carbon dioxide	Human respiration, combustion
Lead	Paints, pipes, outdoor air
Allergens	Moist conditions, outside air
Chemicals	Cleaners, coatings

SOURCE: "Indoor Pollutants," National Academy of Sciences (1981).

tion causing humidity, temperature, and air circulation problems. The remaining 20 percent of the problems are due to miscellaneous causes.

Typical occupant symptoms signifying indoor air problems are eye irritation, sinus congestion, headaches, fatigue, difficulty in wearing contact lenses, itchy skin, sneezing, nausea, and dizziness. The symptoms are due to the contaminants entering the body through the skin, by eating and drinking, or by breathing the contaminated air.

Airborne particles can be controlled through filters. Primary particle types are:

- *Dusts* are mechanically generated solid particles. Over 10 microns in size, they collect on room surfaces and pose a problem of cleanliness. Below 10 microns in size, they can be inhaled into the human system.
- *Smokes* are small solid or liquid particles produced by incomplete combustion and are one of the finest forms of dust.
- *Fumes* are solid particles formed by condensation of vapors of solid materials.
- *Pollens* are moderate sized organic particles that may be a source of irritation for some allergic individuals.
- *Mists and fogs* are suspended liquid particulates.
- *Vapors and gases* are another state of air contaminants.

By far the worst contributor to poor indoor air quality is environmental tobacco smoke. Over 4000 different chemical species have been isolated from tobacco combustion products. Carbon monoxide and respirable particulate matter are major products.

Radioactive materials, particularly radon, are the second most critical contributors and require special monitoring devices. Control must be achieved by preventing the contaminants from getting inside the environment or by collecting the contaminated air as close to the source as possible. The contaminant should be removed with as high an efficiency as possible before releasing it to the outdoors.

Mean Radiant Temperature

The body exchanges heat with the ambient air by convection and conduction and with the surrounding surfaces by longwave radiation. As indicated in figure 3-1, radiation is an important factor in heat transfer at lower temperatures. Body heat losses due to radiation depend primarily on the temperature of surrounding wall, floor, and ceiling surfaces. The more these surface temperatures vary from the average skin temperature, the greater is the radiant heat loss (or gain).

The *mean radiant temperature* (MRT) is a weighted average of the various radiant heat influences in the space. These surfaces and energy sources are weighted according to the area of the thermal influence that is exposed to (or *seen* by) the occupant's body surface.

$$\text{MRT} = \frac{A_1 T_1 + A_2 T_2 + \dots A_N T_N}{A_1 + A_2 \dots A_N}$$

where A_N is the projected area of a specific surface or object and T_N is the temperature of that surface.

Locally variable thermal conditions may occur when the occupant is positioned near a large-area surface that is significantly warmer or cooler than the average surface temperature of the space. Such conditions exist when an individual is sitting adjacent to a single pane large glass window on a cold winter day in a heated interior. Underground spaces provide these conditions by having relatively cold wall and floor surfaces which are in contact with the earth while the ambient air is at a comfortable temperature.

Surface radiant heating systems must be limited in their temperature extremes so as not to exceed overall thermal comfort ranges. Floors should not exceed 85°F nor should ceilings exceed 115°F surface temperatures.

Compensating Effects of Surface Temperature

Although radiant heat transfer is not affected by the other factors of thermal transfer, there is a subjective relationship between thermal comfort and MRT. MRT can be used to compensate for other factors in adjusting the thermal comfort zone. For example, as the cooling or heating potential of environmental air is increased or decreased, the effect can be somewhat offset or canceled by an *inverse* change in room surface temperatures.

In this regard, the psychrometric chart comfort zone in figure 3-2 is based on the condition that the room surface temperatures are at or near the environmental air temperatures (that is, MRT is near the dry bulb temperature). If room surface temperatures are *cooler* than this, comfort will normally occur at higher air temperatures than is indicated in the diagram. On the average, a change of one degree F in the MRT is equivalent to about 0.75°F in the air temperature.

REFERENCES

American Society of Heating, Refrigerating and Air-Conditioning Engineers, Inc. (ASHRAE). 1987 Handbook: *HVAC Systems and Applications;* 1989 Handbook: *Fundamentals;* Standard 62-1989: *Ventilation for Acceptable Indoor Air Quality;* Standard 55-1981: *Thermal Environmental Conditions for Human Occupancy.* Atlanta.

CHAPTER 4

Transitional Patterns

The experience of space is a dynamic one, with periodic or constant occupant movement between areas. When an occupant moves to an adjacent space, his orientation will change; and he may, at some point, become aware of the dominance of a new environment. This transition can therefore be developed to provide a sense of *continuity,* in the sense that luminous, thermal, and sonic influences are similar in the two adjacent spaces. Or, the transition can be developed to provide a sense of *environmental contrast* and change.

To carry this idea of environmental contrast a step further, we observe that when an individual moves from a space that is seriously deficient in light, heat, or sound into a space where these influences are noticeably greater in intensity, the new environment is generally considered to be more suitable and comfortable. But if, in successive steps, the various stimuli continue to increase in intensity, we find that at some point the positive influence wanes, and the same stimuli tend to become negative influences in the environment.

Figure 4-1 illustrates this relationship and shows that when each stimulus is evaluated as an influence on human experience and comfort, there is: a *positive* range of experience in which the occupant is at least momentarily conscious of an improvement in his environment; a *neutral* (comfort) range in which the occupant is free from the negative impingement of environmental deficiencies or excesses; and a *negative* range of experience in which the occupant is at least momentarily conscious of a deterioration in his environment.

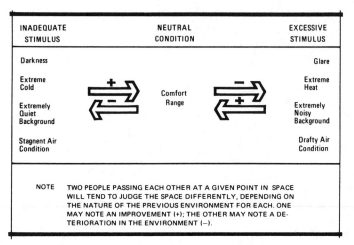

INADEQUATE STIMULUS	NEUTRAL CONDITION	EXCESSIVE STIMULUS
Darkness		Glare
Extreme Cold	Comfort Range	Extreme Heat
Extremely Quiet Background		Extremely Noisy Background
Stagnent Air Condition		Drafty Air Condition

NOTE   TWO PEOPLE PASSING EACH OTHER AT A GIVEN POINT IN SPACE WILL TEND TO JUDGE THE SPACE DIFFERENTLY, DEPENDING ON THE NATURE OF THE PREVIOUS ENVIRONMENT FOR EACH. ONE MAY NOTE AN IMPROVEMENT (+); THE OTHER MAY NOTE A DETERIORATION IN THE ENVIRONMENT (−).

Figure 4-1. Transitional adaptation (typical behavioral patterns).

This diagram also indicates that when transient occupants are involved, these positive or negative interpretations will not be the same for individuals who are moving from different stages of adaptation. It is obviously important, then, for the designer to make some estimate of the individual's relative position in space—for subjective judgment of light, heat, or sound will be affected by the reference level established by the immediately preceding environmental condition.

ADAPTATION

Adaptation refers to the characteristic action of the human body which causes it to seek a state of equilibrium and enables it physiologically to adjust to the prevailing environmental conditions. In essence, after a short period of adjustment to a static environmental stimulus (light, heat, or sound), adaptation tends to neutralize the occupant's sensitivity to the intensity or magnitude of that stimulus. Once this adjustment has taken place, however, this state of individual adaptation may tend to intensify sensitivity to a new change.

When anticipating the effect of adjacent or successive environmental conditions, then, the designer must consider the probable temporary physiological condition of the senses: the degree to which the iris of the eye is open or closed; the degree to which various color receptors (cones) are fatigued or overly sensitive; the level of background noise to which the occupant has become accommodated; and the condition of the skin and the probable effect of changes in temperature, humidity, and air motion.

The occupant's subjective response to a new environmental condition will be affected by the imbalances that the new space creates for any of these senses—and the degree of change involved.

Abrupt Change or Environmental Contrast

Abrupt changes in the intensity of a given stimulus can exert a temporary but important influence on the attitude of the occupant. For example, a noticeable change in light, a significant change in air flow near the occupant, or an increase in background noise may each cause the occupant to become conscious of that particular sensory influence.

If the occupant is subjected to extremely low rates of information input, he may react with interest or a sense of relief from boredom. In this situation, then, the abrupt change is a useful and stimulating influence. On the other hand, the occupant's reaction may also tend toward increased irritation and a feeling of tension when he is involved in a highly complex task or one that requires intensive vigilance. For this reason, when the occupant is subjected to more complex activities or tasks, environmental variations that affect adaptation should be avoided in favor of a more neutral condition.

Transitional Influences Related to Brightness

Visual transitions occur during movement between areas throughout the building. While spatial continuity is often appropriate and necessary, subtle contrast can become a useful source of stimulation and visual relief for an individual who moves about, but remains within the building for extended periods of time.

Furthermore, the casual occupant's patterns of attention and circulation may be influenced by his visual attraction to areas of prominent visual contrast. Lacking other incentives, the individual will generally tend to move toward an area of higher brightness.

However, since perception of brightness intensity varies subjectively with the adaptation of the individual eye, it is necessary that transitional measurements be interpreted in this sense. For this purpose, there have been several attempts to develop scales of *apparent* or subjective brightness. And while these efforts must still be regarded as limited in accuracy, they are nevertheless adequate and useful for general estimating purposes.

Figure 4-2 shows the results of one series of studies. The diagram relates apparent changes in subjective magnitude (vertical scale) to actual physical changes in intensity (horizontal scale) for several levels of eye adaptation.

Transitional Influences Related to Color

After a period of visual concentration on a small bright form or area, an *after-image* of the same shape and complementary hue will appear for a few

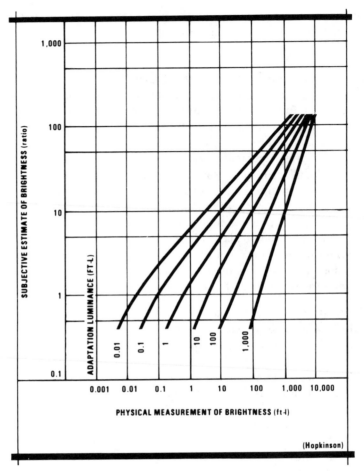

Figure 4-2. Adaptation and subjective brightness.

seconds. This image is due to temporary overstimulation and fatigue of some of the cones in the retina of the eye. Duration of the fatigue, and therefore the duration and strength of the image, will depend on the brightness of the original form and the length of time it was viewed.

Successive Contrast
    Successive contrast is related to this phenomenon. This effect produces a temporary change in the perception of color as the eye moves (after a period of exposure) from one colored surface to another. The perceptual shift is again due to a partial fatigue of some of the cones, increasing the relative sensitivity of the eye to wavelengths *complementary* to the original color.

Successive contrast is also observed after exposure to tinted or saturated colors of light. This action tends to exaggerate spatial color shifts as an individual moves between spaces that are lighted with sources of differing *whiteness.* For example, a visually *warm* space will initially appear much warmer to an individual entering from a visually *cool* space than it will to an individual entering from another *warm* space.

Simultaneous Contrast

Simultaneous contrast is another related visual phenomenon that occurs when a neutral surface is surrounded by color. The neutral panel (particularly white or gray) will appear to be tinted a color *complementary* to the background.

Two or more adjacent colored surfaces viewed simultaneously will affect each other in a similar way; and simultaneous contrast may alter the apparent hue, saturation, and brightness of wall, floor, or ceiling surfaces. However, it should be noted that this effect is significantly reduced when the adjacent areas of color are separated by a narrow neutral area (white, beige, gray, black).

Transitional Influences Related to Sound

A change of 3 dB in background noise intensity will tend to be barely perceptible. The change will be clearly perceptible if it exceeds 7 dB.

Changes in reverberation time are also perceptible if the occupant moves from a very soft, absorbent space to a hard space (or vice versa). Generally, the change in reverberation time between spaces must approximate 0.75 sec or more in order to be a significant subjective change.

(The effects of intermittent variations in white noise are discussed in chapter 2.)

Transitional Influences Related to Temperature and Air Flow

Abrupt changes in environmental temperature or air flow in the vicinity of the body will both tend to cause an immediate increase or decrease in skin temperature. As a general rule such transitional changes tend to be pleasant when they involve noticeable increases in air flow (above the 50-100 ft/min range, but below velocities that will be considered *windy*), with the moving air maintained at or slightly above the temperature of the principal air mass in the room.

When air velocity conditions are stable, there are preliminary indications that an environmental change must approximate 3° ET or 6° MRT in order for the change to be clearly perceptible. This conclusion appears to apply when the general thermal environment is near a condition of optimum comfort.

After a short period of exposure, the body will adjust its blood flow to compensate for the new condition, and the new environment tends to become the *norm.*

Slow Change or Subliminal Contrast

When environmental conditions change very slowly, the rate of change may be *subliminal* (that is, below the threshold of sensation). When this occurs, the body mechanism is tending to *adapt* or adjust itself to accommodate each successive change, making the subjective effect almost imperceptible. This condition occurs both when the stimulus is increasing in intensity and when it is declining.

As the occupant moves from a bright space to a dim one, therefore, he is immediately conscious of a major change in the sensory environment. However, if a room becomes visually *dim* (or thermally *stuffy*) quite gradually, the occupant may not perceive the change, unless he is exposed to an abrupt renewal of the original condition.

When environmental continuity is required, changes in the intensity of various stimuli should be subliminal.

Perceptible Differences in Stimulus Magnitude

Observations and experiments indicate that minimum perceptible differences in stimulus are not constant intensities. Instead, a *just noticeable difference* is found to be a *ratio* (as described in Weber's Law and the Weber-Fechner Law).

$$\frac{M_2}{M_1} = K,$$

where M_1 is the magnitude of the original stimulus; M_2 is the magnitude of the new stimulus; and K is the applicable minimum ratio (constant).

While this system of ratio relationships is not universally true (particularly in regard to thermal stimuli), it is approximately true in reference to vision, hearing, pressure, and muscular action. In each case, the rule is most accurate in the middle regions of the sensory scale, and becomes more uncertain near the upper and lower limits of perception. So although the principle lacks precision and authority in some respects, it nevertheless serves as a valuable approximation in many phases of environmental design.

The precise minimum ratio will vary with the sense being measured. But for brightness and sound intensity, the approximate minimum meaningful change in intensity is a ratio of two. As a result, general brightness differences between adjacent spaces of less than 2:1 tend to be perceptually insignificant. This means that successive doubling of brightness intensity is the minimum ratio that will produce a sensation of perceptible steps of change.

Similarly, a meaningful change in sonic intensity requires at least a doubling of power level. On the logarithmic decibel scale, this is a minimum change of 3 dB (see table 6-1).

Thermal influences are not generally subject to the *constant ratio* rule. However, changes of 3° ET or 6° MRT have been previously identified as perceptible near the comfort zone. Short-term changes of lesser magnitude tend to be subliminal.

PART TWO

SYSTEM PERFORMANCE AND METHOD: PERFORMANCE OF MATERIALS, FORMS, AND INDIVIDUAL SYSTEMS

Sooner or later, discussions of environmental design must move beyond the evaluation of objective and subjective environmental criteria to the selection of techniques to achieve the intended result. This phase involves equipment and material selection, assembly and installation, and the need to assimilate and coordinate the physical components into a logical and harmonious building system.

In this regard the designer is guided by a sense of comprehensive integrity and system logic that must dominate the selection and use of environmental devices. For example, a beam of light depends on the integrity of the relationship between a light source, a reflector, and a louver, baffle, or lensing medium. A change in any one of these will alter the characteristics of the beam of light, affecting one or more of the objective and/or subjective criteria of the luminous environment.

Visual perception of a space (in terms of brightness, form, or color) depends on the integrity of the relationship between direct and reflecting lighting elements, specific surface finishes, and the subjective reference level of the observer. A change in any one of these will alter perception.

A sense of acoustical quality depends on the integrity of the relationship between workstation space dividers, the ceiling, floor, and perimeter walls; the nature of surfaces and finishes; density of room occupancy; and the source-to-background signal strength. A change in any one of these will alter sonic perception.

An individual's perception of thermal comfort depends on the relationship between various heat-producing and heat-consuming elements in a space.

In the recent past, architects and engineers have accomplished design goals by relying heavily on systems using large quantities of fossil-fueled energy. A more energy conservative approach, and one which may lead to a more human-oriented solution, is the balancing of *natural* forces such as sun, wind, and water within the building zone. The responsible designer will seek a marriage between natural and mechanical energy that satisfactorily provides for human functions.

Designers must be conversant in the use of meaningful devices, materials and forms, as well as in the manipulation of their relationships. These are fundamental in any attempt to develop a successful sensory environment.

CHAPTER 5

Light Generation and Control

The discussion of the performance parameters of the luminous environment in chapter 1 indicated the scope and number of factors to be considered for successful lighting design. Singling out only one or two of these performance factors and selecting equipment and equipment locations accordingly was a common practice during the 1950s, 1960s, and early 1970s. Particular emphasis was placed on illuminance (footcandle) levels which became the primary criterion for lighting system selection in that era of "more light, better sight." Now, and through the last decade, power budgeting has replaced illuminance as the primary driving force. Such short-sightedness may result in environments exhibiting low power loads but with little or no mood–atmosphere nor composition and providing little human motivation. This situation may result in lost human interest and productivity, and subsequently in increased energy consumption in order to get the job done.

This chapter discusses the many tools available to achieving the performance standards established in chapter 1. There is no panacea. The methods that are discussed may appear to be attractive means to achieving one performance criterion, yet may offset other criteria. The competent designer will be able to assemble and integrate a host of solutions and systems to *best* achieve effective visual environment performance standards.

VISIBLE LIGHT

The human environment is subject to a number of natural and man-made sources that emit energy in various regions of the electromagnetic spectrum (fig. 5-1). The segment approximately defined as 380 to 760 nanometers is generally referred to as the *visible spectrum* (the band of energy to which the

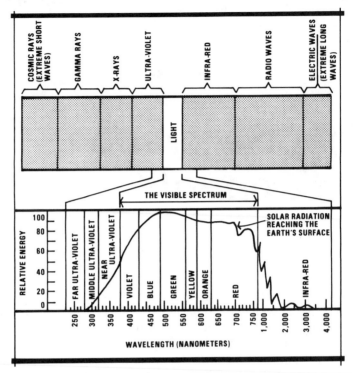

Figure 5-1. The electromagnetic spectrum.

human eye is sensitive). The term *light* refers to energy emitted in this region. Since light is required for visual work, orientation, and spatial definition and circulation, spaces for human occupancy must provide for these by introducing either natural or electric light sources.

DAYLIGHTING

The sun is of course the dominant lighting influence in nature. All other natural lighting influences and forms perform in response to this source. The sun emits a continuous spectrum of energy, and because of the very high source temperature involved, the total spectrum ranges into the longer and shorter wavelengths on both sides of the visible band. Intensity of energy throughout the visible range is fairly uniform.

The atmosphere that surrounds the earth, however, modifies solar energy by absorption, reflection, and selective scattering. This action tends to limit the direct penetration of energy in both infrared and ultraviolet regions. It also tends to modify the intensity at various wavelengths in the visible region. The water vapor content of the atmosphere is particularly important, as this produces transmission changes ranging from clear sky to hazy and overcast conditions. These transient characteristics of the atmosphere tend to produce extremely variable light conditions.

As a further variable, the tilt (declination) of the earth's axis and normal daily rotation combine to produce a constantly changing angular relationship between the sun and any specific location on earth. This, in turn, produces changes in solar intensity on both an hourly and seasonal basis.

Daylight Analysis

In analyzing natural light at a given location, there are three basic components to be considered: direct sunlight, which impinges intermittently on the east, south, or west exposures of a building in the northern hemisphere; skylight, which impinges simultaneously and somewhat more consistently on all exposures of a building; and reflected light from the ground and from nearby man-made elements. Each of these components will vary with time of day, season, and prevalent atmospheric conditions.

Direct Sunlight

Direct sunlight is coincident with maximum solar heat. For this reason it is generally desirable to control the penetration of this component into the interior, or at least selectively filter infrared solar radiation with glazing. When developing shielding devices similar to those shown in figure 5-2 or where limited penetration may be desirable for occasional spatial effect, such as that shown in figure 5-3, the action of sunlight can be predicted through graphic projection of solar rays, using azimuth and altitude angles. Typical angles are listed in tables 5-1, 5-2, and 5-3, with a complete listing in the *ASHRAE* (American Society of Heating, Refrigerating and Air-Conditioning Engineers) *Handbook*. Typical illuminances to be expected from direct solar exposure are also listed in tables 5-1, 5-2, and 5-3.

Skylight

Although variable in intensity and color, skylight tends to be a more consistent source of natural daylight than direct sunlight. Skylight, however, must be evaluated in at least two different forms: the overcast sky condition and the clear sky condition. These are similar in that they represent a large-area diffuse light source (as contrasted with the sun, which functions environmentally as a high intensity, small-area or point source). The two forms of skylight differ in intensity, color quality, and directional uniformity (see tables 5-1, 5-2, and 5-3). The method for estimating the relative intensity of skylight on vertical and horizontal building surfaces is discussed in note 2 in table 5-1.

Figure 5-2. Vertical architectural baffles shield the windows from direct solar exposure.

Figure 5-3. Narrow skylights at the building roofline permit daylight beaming.

Table 5-1. Typical daylight design data—1.

December 22 (40° North Latitude)

	Solar Time	Location of Sun	
		Azimuth[1]	Altitude
Sunrise	7:30	121°0′	0°0′
	8:00	127°0′	5°30′
	Noon	180°0′	26°30′
	4:00	127°0′	5°30′
Sunset	4:30	121°0′	0°0′

1. *Azimuth* refers to the horizontal angle from north.

	Approximate Average Sky Brightness (footLamberts)		
	8 AM	Noon	4 PM
Typical overcast condition	190	930[2]	190
Typical clear day condition			
N	275	525	275
E	975	875	275
S	750	2,175	750
W	275	875	975

2. A uniformly bright sky of 930 fL-l would produce approx. 930 fc on the ground and on all horizontal surfaces.
 Vertical illumination from this source is approximately:
  Sky effect: Sky footLambert × 0.50.
  Ground effect: Sky footLambert × 0.50 × ground reflection.

	Typical Direct Solar Illumination (footcandles)		
	8 AM	Noon	4 PM
Perpendicular[3] to sun rays	2,250	6,650	2,250
Horizontal	100	2,850	100

3. Illumination on a given plane is obtained by multiplying perpendicular illumination times the cosine of the angle of incidence.

Reflected Light from the Ground

The method for estimating the relative intensity of ground light on vertical building surfaces is also shown in table 5-1 (note 2). Since the ground surfaces are subject to control by the architect, the selection of adjacent landscaping materials becomes a moderate control device to influence the intensity of daylight that is incident on the building shell (table 5-4). Further-

Table 5-2. Typical daylight design data—2.

March 21, September 21 (40° North Latitude)

	Solar Time	Location of Sun	
		Azimuth[1]	Altitude
Sunrise	6:00	90°0'	0°0'
	8:00	110°30'	22°30'
	Noon	180°0'	50°0'
	4:00	110°30'	22°30'
Sunset	6:00	90°0'	0°0'

1. *Azimuth* refers to the horizontal angle from north.

	Approximate Average Sky Brightness (footLamberts)		
	8 AM	*Noon*	*4 PM*
Typical overcast condition	790	1,760[2]	790
Typical clear day conditions			
N	725	850	725
E	2,450	1,475	600
S	1,700	2,700	1,700
W	600	1,475	2,450

2. A uniformly bright sky of 1,760 fL-l would produce approx. 1,760 fc on the ground and on all horizontal surfaces.
  Vertical illumination from this source is approximately:
    Sky effect: Sky footLambert × 0.50.
    Ground effect: Sky footLambert × 0.50 × ground reflection.

	Typical Direct Solar Illumination (footcandles)		
	8 AM	*Noon*	*4 PM*
Perpendicular[3] to sun rays	6,050	8,250	6,050
Horizontal	2,050	6,000	2,050

3. Illumination on a given plane is obtained by multiplying perpendicular illumination times the cosine of the angle of incidence.

more, since much of the light that is reflected from the ground is directed toward the interior ceiling surface, the interior distribution and intensity of daylight in low buildings can be manipulated through the interaction of these two reflecting surfaces (ground and ceiling).

For low buildings on a sunny day, light reflected from the ground will typically approach 50 percent of the total daylight incident on a window area

Table 5-3. Typical daylight design data—3.

June 21 (40° North Latitude)

	Solar Time	Location of Sun	
		Azimuth[1]	Altitude
Sunrise	4:30	59°0'	0°0'
	8:00	89°0'	37°30'
	Noon	180°0'	73°30'
	4:00	89°0'	37°30'
Sunset	7:30	59°0'	0°0'

1. *Azimuth* refers to the horizontal angle from north.

	Approximate Average Sky Brightness (footLamberts)		
	8 AM	Noon	4 PM
Typical overcast condition	1,290	3,060[2]	1,290
Typical clear day conditions			
N	1,325	950	1,325
E	2,850	1,400	700
S	1,350	2,400	1,350
W	700	1,400	2,850

2. A uniformly bright sky of 3,060 fL-l would produce approx. 3,060 fc on the ground and on all horizontal surfaces.
   Vertical illumination from this source is approximately:
   Sky effect: Sky footLambert × 0.50.
   Ground effect: Sky footLambert × 0.50 × ground reflection.

	Typical Direct Solar Illumination (footcandles)		
	8 AM	Noon	4 PM
Perpendicular[3] to sun rays	7,600	8,850	7,600
Horizontal	4,550	8,100	4,550

3. Illumination on a given plane is obtained by multiplying perpendicular illumination times the cosine of the angle of incidence.

that is shaded from direct sunlight. It may exceed this proportion when the building is surrounded by high reflectance surfaces, such as sand, light concrete, or snow cover. For overcast conditions the light reflected from the ground will be less significant, typically representing approximately 10 to 25 percent of the daylight incident on the window. Figure 5-4 illustrates a typical daylighting analysis for a small building (or room of a larger building).

Table 5-4. Typical daylight design data (ground reflectance).

	Typical Reflectance (%)
Grass	6
Vegetation (mean value)	25
Earth	7
Snow	
new	75
old	60
Concrete	55
White marble	45
Brick	
buff	45
red	30
Gravel	13
Asphalt	7
Painted surface	
new white	75
old white	55

Several different sky conditions and ground reflectances were used to indicate their influence on daylighting in buildings.

Man-Made Enclosures

Throughout history, building design has involved a paradox. There is the need to provide structures that are adequately sealed to prevent the intrusion of adverse influences such as cold winds and rain, yet these same structures must be adequately open to permit the penetration of light and ventilation. Each society and culture has produced its own solutions to this problem, reflecting the specific demands of the region, the technology of the time, and the sensitivity of the people.

Early Western civilizations developed in the arid and semiarid regions near the 70 degree isotherm. In these regions, the natural lighting condition is both intense and seasonally consistent. Minimal openings in Egyptian and Greek structures were often sufficient to permit adequate light to penetrate into the interior for purposes of spatial definition and orientation. Similarly, exterior sculptural detail and building form were relatively subtle because intense, directional light is sufficient to produce sharp shadows for contrast.

In medieval Europe, Gothic and Baroque architecture evolved more open structures to facilitate penetrations of the more gentle and variable daylight

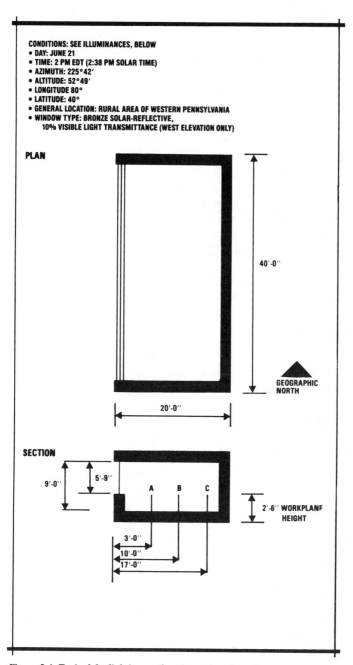

CONDITIONS: SEE ILLUMINANCES, BELOW
- DAY: JUNE 21
- TIME: 2 PM EDT (2:38 PM SOLAR TIME)
- AZIMUTH: 225°42'
- ALTITUDE: 52°49'
- LONGITUDE 80°
- LATITUDE: 40°
- GENERAL LOCATION: RURAL AREA OF WESTERN PENNSYLVANIA
- WINDOW TYPE: BRONZE SOLAR-REFLECTIVE,
     10% VISIBLE LIGHT TRANSMITTANCE (WEST ELEVATION ONLY)

PLAN

40'-0''

GEOGRAPHIC
NORTH

20'-0''

SECTION

9'-0''

5'-9''

A   B   C

2'-6'' WORKPLANE
HEIGHT

3'-0''
10'-0''
17'-0''

Figure 5-4. Typical daylighting performance. (continued)

PROGRAM	CONDITIONS	LOCATION A	B	C
DAY 01	• REFLECTANCES (CEILING, WALLS, FLOOR) 80%-50%-20% • CLOUDY • GROUND REFLECTANCE OF 6%	42	11	6
DAY 02	• SAME AS DAY 01 EXCEPT CLEAR	765	43	31
DAY 03	• SAME AS DAY 01 EXCEPT NO GROUND REFLECTANCE	41	10	6
DAY 04	• SAME AS DAY 02 EXCEPT NO GROUND REFLECTANCE	759	35	29
DAY 05	• SAME AS DAY 01 EXCEPT GROUND REFLECTANCE OF 55% (E.G. CONCRETE)	51	16	10
DAY 06	• SAME AS DAY 01 EXCEPT WALL REFLECTANCES OF 70% AND FLOOR REFLECTANCE OF 30%	44	13	9

NOTE: CALCULATIONS WERE PERFORMED WITH THE LUMEN II
LIGHTING PROGRAM.

Figure 5-4. (continued). Illuminances (average footcandles on workplane due to daylight).

that is associated with temperate climates. At the same time, stonework tracery was developed to provide more richly molded forms because natural shadows in these conditions tend to be more subtle.

Buildings in the temperate zones remain subject to wide variation in the character of daylight—particularly in the northern and middle areas of the United States where diffuse daylight conditions tend to predominate during a majority of the days in a year. The following comments relate to buildings located in north temperate areas. They summarize some of the more basic physical relationships involved in using daylight in this region.

Orientation of Building Faces

Assuming suitable sun control, the south face of the building generally affords the maximum quantities of light. This is particularly true during the winter months (see tables 5-1, 5-2, and 5-3). Because of the low angles of the sun during the morning and afternoon hours, the east and west exposures present the most difficult problems related to direct sunlight penetration and glare control. These same exposure conditions tend to complicate the thermal environment as well.

Except for very short early morning and late afternoon intervals during the peak of the summer solstice, the sun does not impinge on the north face of buildings in the northern temperate regions. For this reason skylight and reflected light from the ground are the dominant sources of north daylight; this light remains relatively consistent throughout the day. North exposure is useful, then, when consistency of daylight color and shadow effects is required.

Development of Building Form

Since light intensity diminishes with distance from the source, the penetration capability of daylight is limited. For this reason a decision to use daylight will tend to limit the building form to those configurations that permit the introduction of light openings near the significant interior task centers. This may lead to relatively narrow building configurations that permit light to enter from either or both sides; it may lead to low buildings that permit light to enter through roof openings. It may also lead to the use of courtyards and light wells to facilitate the penetration of daylight.

Architectural Daylighting

Outlined on the next several pages are a variety of methods for introducing daylight into the built environment. Recognize that daylight may penetrate through either image-preserving (transparent) or non-image-preserving (translucent) media. Two properties of these media that relate to daylighting design are visible light transmittance from outside to inside and visible light reflectance of the inside surface.

Visible light transmittance from outside to inside is necessary for determining the quantity and quality of light entering the space. This transmittance is usually given as a percentage. A window, for example, of the solar-reflective genre may have a total visible light transmittance of eight percent, indicating only eight percent of the daylight striking the window is transmitted to its interior surface. As sophisticated daylight schemes evolve that attempt to maximize both thermal and daylighting benefits, transmittant "sandwiches" of materials may be developed. Visible light transmittance of such "sandwiches" must be obtained empirically, because calculation of such transmittance values is misleading due to the variety of interreflections occurring between the materials within the sandwich.

The designer must establish appropriate criteria for daylighting. There may be instances where glimpses of an exterior view are the sole "visual" criterion related to daylighting. In this case small, image-preserving windows may be used with negligible daylighting contribution, while contributing to a good building-skin thermal performance. Large area daylight contributions, however, can also be responsible for good building energy efficiency with appropriate attention given to controls and to *spectral transmission char-*

acteristics of the glazing. The key to successful daylighting in most climates is automatic controls. Manual systems permit the occupants either to forget about or consciously neglect taking action. Automated systems, therefore, are generally used if consistent and reliable building energy optimization is desired. Controls may include stepped switching, continuous dimming, or simply zoned switching. These systems require strategically located photocells to determine daylight levels and signal the controls to increase or decrease electric lighting levels. Dimming systems are particularly desirable since electric light levels can be changed at a rate which goes unnoticed by occupants, eliminating distracting brightness shifts. Also, within the past several years, the dimming systems for fluorescent lighting have become more cost effective as material costs have dropped. Particularly promising is electronic dimming which has additional benefits of no flicker and no audible noise or hum.

The use of controls and the extent of their sophistication depend upon the method or methods of introducing daylight into the space. Architectural daylighting can be introduced by windows, skylights, clerestories, courtyards, atria, light shafts and sunbeaming, and translucent roof membranes. Daylight contribution to interior spaces is not only influenced by the type of architectural feature used, but also by surrounding building sizes and their proximity and finishes; ground finishes; sky conditions (clear, cloudy, partly cloudy); pollution conditions (industrial, urban, or suburban siting); orientation (north, south, east, west); time of year; time of day; and latitude and longitude.

Window Elements

Windows can provide building occupants with a view of the outdoor environment, and need not be large to accomplish this goal. This external-view function has a significant influence on the occupant's sense of orientation, environmental relationship, and well-being. Non-image-preserving window treatments suppress this influence.

Windows also function as a source of illumination during the daytime. The quantity of light introduced into a space by windows depends on the window size, transmittance, and proximity to the ceiling and/or task.

When a sense of general brightness consistency (uniformity) is desired from windows, the effective depth (width) of rooms in more northern temperate regions where there are a substantial number of overcast days should be limited to a maximum of two and one-half times the distance from the floor to the window head (fig. 5-5). This assumes the use of continuous or near-continuous window elements.

In regions where clear days are more prevalent and reliable, general brightness consistency exists over a zone of three and one-half times the floor-to-window-head distance. When these ratios are exceeded, brightness

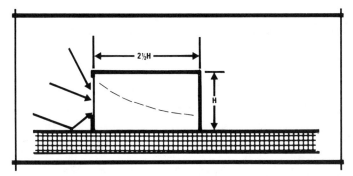

Figure 5-5. Daylight distribution: single window.

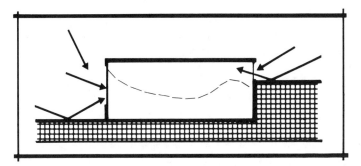

Figure 5-6. Daylight distribution: opposite window walls.

in the interior zones of the space will be noticeably lower than in perimeter zones. This usually causes some sort of user reaction, including switching "on" electric lighting, moving toward perimeter zones, or performing work more slowly—not necessarily because of limited illumination, but because of the distracting bright surrounds. Lightshelves can be effective in balancing room surface brightnesses and more uniformly distributing daylight throughout a space. The top surface of the lightshelf must be high reflectance, spectrally neutral and matte in finish. Recognize that low reflectance surfaces will be inefficient, negating the potential benefit of the lightshelf. Neutral white surfaces are necessary to avoid a color band of light on the ceiling. Matte finishes will uniformly reflect daylight across the ceiling surface, while specular finishes will cause brightness bands resulting in glare and nonuniformity.

When the previously discussed room width-to-window height ratios cannot be maintained, one option that will permit the development of wider room sections (for a given ceiling height) is the addition of window openings in the opposite wall (see fig. 5-6). This may be a full or partial window section, and may also contribute to the natural ventilation of the space by providing cross-ventilation—with operable sashes. Since at least one of the window elements may face the sun during significant portions of the day, some type of sun control device may be required.

Generally, the most significant single interior surface for assisting the interior reflection of daylight is the ceiling. When improved penetration and improved spatial brightness uniformity are desired, the ceiling should have a diffuse, high-reflectance finish (70 percent to 90 percent). High-reflectance floor finishes (25 percent to 35 percent) are also significant in assisting the interreflection of light. In any case, surfaces should be matte. Specular (shiny) or even semispecular surfaces can cause reflected glare.

The development of sophisticated schemes to allow the further penetration of daylight into the interior regions of buildings through windows is an important function of the 1980s building evolution period. The designer should recognize, however, that specular reflectors are generally less successful than diffuse reflectors unless sun-tracking is undertaken. Even then, the severe brightness ratios exhibited within the space may make specular reflection of sunlight undesirable unless light pipe lensing media are used to diffuse or spread the light more uniformly. High technology materials and electronics may soon make this daylight method feasible, both economically and physically.

Openings that are intended to facilitate the penetration of daylight are also likely to become sources of sky glare and sun glare. This may necessitate the use of a brightness control device such as external baffles, image-preserving roll shades, venetian blinds, glare-reducing glass, draperies, or tree landscaping. The resulting screen action will, of course, also tend to restrict the penetration and intensity of light in the interior.

Clerestories and Roof Monitors

Clerestories and roof monitors (fig. 5-7) are one category of forms that may be useful for overcoming the previously discussed limitations in the room width-to-window height ratio. Specifically, these devices provide a means for emitting light into the more remote interior portions of spaces that are located immediately below the roof.

Combination Systems

When window walls are supplemented by clerestory windows both openings should be oriented in the same direction (fig. 5-8). The setback distance from the window (S_1) should be approximately one to one and one-half times the window height. When continuous or near-continuous windows are involved this setback can be increased to approximately two times the window height.

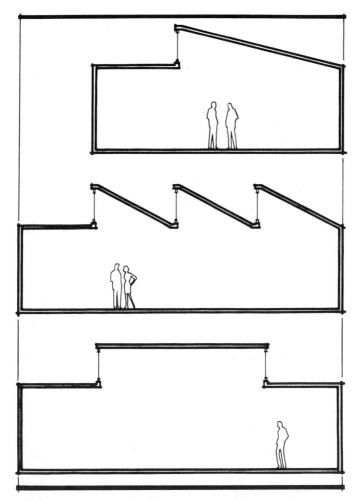

Figure 5-7. Basic clerestory and monitor sections.

When clerestory sill heights are relatively low (close to the floor plane), the height of the clerestory opening should be approximately one-half of the side wall window height ($H_c = $ minimum $\frac{1}{2} \times H_w$). When the clerestory sill is high (approximately two and one-half to three times the side wall window height), the height of the clerestory opening should be approximately equal to the window height. The distance from the clerestory window to the dark far wall (S_2) should not exceed two times the height to the top of the clerestory (H_t). For small clerestory openings (H_c) this distance should be reduced to one and one-half times. When greater-than-manageable S_2 distances are involved consider the roof monitor.

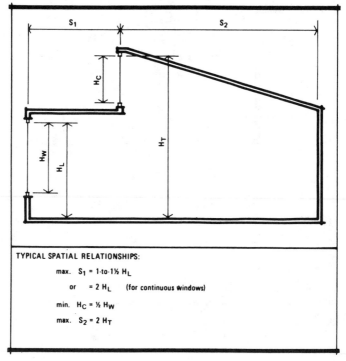

TYPICAL SPATIAL RELATIONSHIPS:

max. $S_1 = 1$-to-$1\frac{1}{2}$ H_L

or $= 2 H_L$ (for continuous windows)

min. $H_C = \frac{1}{2} H_W$

max. $S_2 = 2 H_T$

Figure 5-8. Combination system proportions.

Roof Openings and Skylights

Roof openings can also provide daylighting for deep or isolated interior spaces. The principal disadvantage is the fact that major solar heat loads generally impinge on the horizontal roof surface, and these loads will penetrate directly into the interior space through the openings. This problem of solar penetration is a particularly significant consideration for roof openings and skylights because direct solar loads are incident on the roof continuously over most of the day, rather than for the shorter peak periods associated with individual wall orientations. For this reason louvers, baffles, light wells, external canopies, or insulating shielding devices are sometimes used to reduce direct solar penetration (figs. 5-9 and 5-10).

Regarding the placement of openings, if the system is intended to produce uniform brightness on an interior horizontal surface, moderate-sized openings should be placed so that the center-to-center spacing does not exceed two times the floor-to-ceiling height (or, for work areas, two times the height of the ceiling above the workplane). Small-area roof openings provide a much more narrow distribution. For this reason the maximum center-to-center spacing of smaller openings should not exceed the floor-to-ceiling distance when horizontal brightness uniformity is desired.

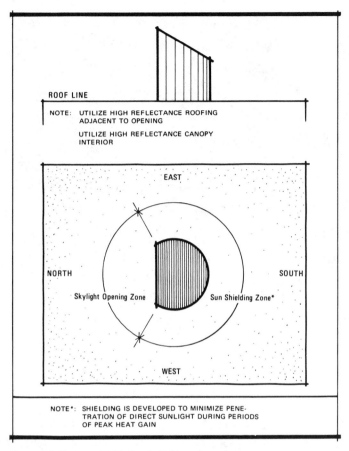

Figure 5-9. Canopy shields for skylight roof openings.

The effectiveness of light wells will vary with the size of the well opening, the depth of the well, and the reflectance of the well walls. These factors can be related with the following formula:

$$I_w = \frac{H_w \times (W_w + L_w)}{2 \times (W_w \times L_w)},$$

where: I_w is the well index (from fig. 5-11); W_w is the width of the well opening; L_w is the length of the well opening; and H_w is the depth of the well opening.

For general estimating purposes, figure 5-11 further relates these factors and indicates the relative transmission efficiency of various well configurations. The transmission of the glazing (including the anticipated effect of dirt accumulation) will further reduce this efficiency.

Figure 5-10. Window blinds on skylights act as louvers, allowing for automated daylight control in this case.

Courtyards and Atria

In very large structures, either in plan dimensions or in height, or in cases where the exterior view is of limited quality, courtyards and atria are plausible light-entry methods. Generally, because of their limited size in comparison to the total building, these devices can only contribute daylight to areas in the immediate vicinity, although a view of a naturally lit area from the surrounding work space is desirable. Courtyard and atrium areas usually are intended to support plant growth. Daylighting alone in narrow courtyards with the major axis running east and west may not support the more lush and flowery growth many people prefer. Similarly in atria closed over with low-transmittant glass, plant growth may be disappointing unless supplementary electric lighting is used. A good variety of plants will survive on 2,000 footcandle-hours of illumination (200 footcandles for 10 hours) per day of a reasonably broad spectrum (white light). It is desirable to produce this illumination on as many plant surfaces as possible, not just on their tops. Plant illuminances and spectral requirements should always be verified with the landscape architect or horticulturist.

Sun Beaming and Light Piping

Perhaps the most powerful and conveniently packaged daylight source and yet the least cost effective and most elusive is the sun itself. Maintaining

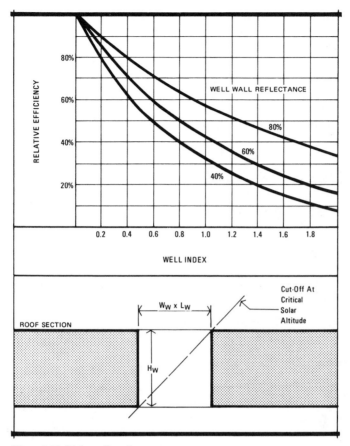

Figure 5-11. Light well performance.

the integrity of the sun beam requires the use of highly polished, specular reflectors. Tracking devices are required to optimize sun beaming over most of the day. At the user end of sun beaming, carefully controlled optical devices are necessary to distribute the beam of sunlight over a task area while avoiding extreme bright-dark striations, which can cause strain, glare, excessive luminance patterns, and distraction. The combination of special reflectors, tracking equipment, and optical assemblies generally is not as cost effective as other methods of daylighting. As new technologies are developed and as fossil fuels become more costly, sun beaming may become much more desirable. The designer should recognize that for particular facilities, such as worship centers and building lobbies, straightforward methods of sunbeaming without special tracking and optic devices may be effectively employed for focal emphasis and mood setting. This is demonstrated in figure 5-3.

One form of sun beaming, which has generated a great deal of interest, is "light piping." A patented version collects sunlight on a building roof and, with optic fiber technology, distributes the light throughout the building. This same concept could be used with high-efficiency electric sources for nighttime and overcast day uses. Although these lighting methods may substantially decrease energy consumption, the designer must remain cognizant of other lighting parameters—luminance, glare, light control, directionality (modeling), mood, psychological responses, and so forth. Figure 5-12

Figure 5-12. Light piping: electric or daylight piping can be used to illuminate large or awkwardly configured internal spaces like this stairwell.

illustrates a light pipe application in a stairwell. During low brightness daylight conditions, electric light is piped through the element.

Translucent Membranes

Another form of daylighting, similar to the skylight, involves using translucent materials as roof membranes. These fabric structures provide an added dimension of architectural design because of the rather free-form roof configurations for spaces where structural elements are unacceptable. For large, column-free interior spaces, like the Pontiac, Michigan, Silverdome shown in figure 5-13, the air-supported fabric membrane is quite desirable. These translucent membranes introduce a uniform, diffuse light into the spaces they cover. Generally, for rather large spaces with very large translucent membrane skylights, the illumination at the floor plane or any other horizontal plane will be about 10 percent (transmittance of the membrane) of the illumination on the roof. For example, on a typical overcast day on March 21, at 40 degrees north latitude at noon (from table 5-2, note 2), 1,760 footcandles will fall on the translucent membrane, then roughly 180 footcandles will fall on horizontal surfaces under the fabric roof.

Electric lighting of these spaces can be difficult, even though the fabric generally has a light reflectance of 70 percent at its inside surface. The fabric is not uniformly diffuse or Lambertian, and therefore does not reflect light equally in *all* directions. Also, because of the free-form nature of some of these roof shapes, carefully controlled aiming is necessary if indirect lighting

Figure 5-13. Air-supported fabric is a lightweight alternative to skylight systems.

is to be used attractively and efficiently. Direct lighting, although used for sparkle and accenting, is generally difficult to mount to the lightweight fabric structure and may not enhance the architectural form as effectively as an indirect approach. The variety of tasks in these structures and their subsequent illuminance requirements may dictate a combination of direct and indirect electric lighting if nighttime and television filming use occurs.

ELECTRIC LIGHTING
Incandescent Lamps

The principle behind incandescent sources is the heating of a metallic solid to a temperature high enough to produce visible radiation. Heated solids will emit all wavelengths of radiation over a certain portion of the electromagnetic spectrum and are therefore known as *continuous spectrum* sources. The radiant energy emitted, however, is not the same intensity at all wavelengths. Furthermore, as the absolute temperature of the solid increases, the energy peak will shift toward the shorter-wave (ultraviolet) end of the spectrum. For example, a solid that is subjected to the normal human environment of 70° to 80°F (530° to 540°K) will generally emit a radiant energy peak at about 10,000 nanometers. Since this energy is not visible to the eye, the solid is considered to be *black*. On the other hand, an incandescent lamp tungsten filament operating at a temperature of 2,900°K (see *Lamp Color,* chapter 1) will peak at about 1,000 nanometers in the near infrared region, with some of the continuous spectrum now being emitted in the visible region below 760 nanometers (fig. 5-14, top).

In the normal range of environmental room temperatures, therefore, objects and surfaces will emit energy in the far infrared region. But as the temperature of the object increases, portions of the visible spectrum are also ultimately emitted. As this action takes place, the human observer first perceives a red glow; as the object temperature continues to increase, the observer perceives a white glow. This relationship of color and temperature is described by the action of a theoretical blackbody radiator as summarized in table 5-5. The performance of tungsten is near the theoretical action shown for the listed *Object Temperatures.*

The *standard* incandescent lamp generally is comprised of a coiled tungsten filament and one of four common lamp enclosures. Coiling the filament provides for a large surface area of tungsten to be exposed, thus being able to produce a fairly large amount of light from a small package. The most common lamp enclosures are the "A," the "G," the "R," and the "PAR."

The A-lamp is a general-service incandescent lamp available in either clear (denoted by "C") or inside frosted (denoted by "IF"). The clear lamps are generally used where specialty luminaire reflectors are employed or used where "sparkle" or "glitter" are appropriate. The shape of the A-lamp,

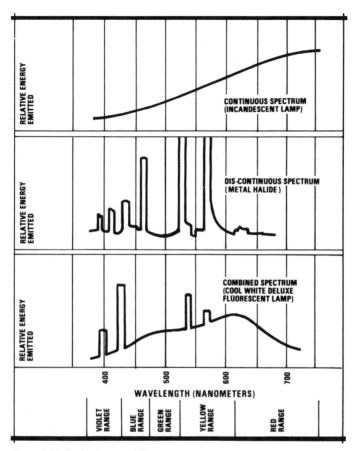

Figure 5-14. Typical spectral diagrams.

Table 5-5. Blackbody radiation.

Object Temperature (°K)	Visible Color Emitted by the Object
Room temperature	Black (no visible emission)
800	Red
3,000	Warm white
5,000	*Crisp white*
8,000	Pale blue
60,000	Deep blue

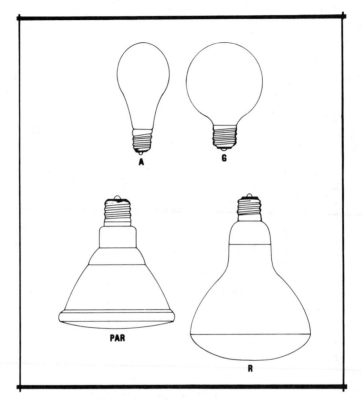

Figure 5-15. Incandescent lamp shapes.

somewhat pearlike, is shown in figure 5-15. The diameter of the large portion of the lamp is given in eighths of inches. A 150-watt A-19 IF lamp designation would signify a 150-watt A-lamp with a diameter of 19/8 inches with an inside frosted enclosure. Figure 5-16 shows A-lamp fitted sconces.

G-lamps are considered the "designer" version of the A-lamp. As shown in figure 5-15, G-lamps are globelike (hence the *G*) in appearance, and are therefore apparently more appealing than the A-lamp when used without shades or luminaire housings. The low-wattage clear G-lamps are used extensively in festive, carnivallike environments.

The A- and G-lamps, when used alone without special reflectors or shades, distribute light rather uniformly in all directions. In an effort to provide a more self-contained directional service lamp, the R (for reflector) and PAR (for parabolic aluminized reflector) lamps were developed. The R-lamp (fig. 5-15) has an opaque silver-reflective coating on the inside of the upper portion of the lamp. This helps direct the light radiating from the filament through the lower portion of the glass bulb. On the PAR lamp (fig.

Figure 5-16. Incandescent lamps: A-lamp wall sconces. Same sconces can be fitted with more efficient, longer life compact fluorescent lamps, if dimming is not a requirement.

5-15), the upper portion of the bulb is actually a parabolic reflector. This reflector produces a narrow light-distribution pattern useful for precise control, such as that required for focus lighting an object. Figure 8-20 illustrates the effect of PAR lamp accents and in the foreground, a PAR lamp downlight. Figure 5-17 shows the effect of PAR lamp uplights in a courtyard.

At times, very special highlighting may require not only very well controlled light but also reasonably high light quantities. Tungsten-halogen (or quartz-iodine) lamps serve these functions. In standard incandescent lamps, in order to produce light, either more filament and/or more electricity is needed. There are limits, however, and as electricity is put through a filament the tungsten will eventually begin to boil away, hence the blackening on the bulb walls of standard incandescent lamps.

In quartz-halogen lamps, iodine causes vaporized tungsten to redeposit onto the filament, permitting a hotter filament which in turn produces more light than a conventional incandescent lamp of equal wattage. Also, the clear quartz envelope (not glass) can accommodate high temperatures and hence can closely surround the filament and provide a very small lamp. The more closely the lamp approaches a point source, the easier it is to control directional qualities and light intensities.

Incandescent lamps are still used extensively in residential interiors and specialty applications in contract interiors. Primary characteristics of incandescent lamps are described in table 5-6. New lamp technologies are emerging which will result in substantially improved efficiencies and life-ratings for

Figure 5-17. Incandescent lamps: PAR-lamp uplights illuminate trees.

incandescentlike sources, further enabling the designer to provide attractive, efficient luminous environments.

Electric Discharge Lamps

Another major method of light production involves energy that is emitted by a gas when that gas is subjected to an electric discharge. There are primarily two classes of electric discharge lamps: "low" pressure and "high" pressure (quotes indicate these terms are used in a relative sense). Presently, the fluorescent class of lamps is the primary low-pressure source acceptable for most interior living and working environments. The high-pressure lamps, generically known as high intensity discharge (HID) sources, are acceptable for many interior environment applications when the visual environments are properly designed.

Fluorescent lamps initially emit energy almost completely in the ultraviolet portion of the spectrum. This is because of the low mercury gas pressure (approximately 1/100,000 of an atmosphere). To become a useful lighting device, the fluorescent lamp must rely on the interaction of the emitted ultraviolet radiation with phosphors—substances that absorb ultraviolet light, resulting in emission of visible light. Depending on the characteristics of the phosphor, certain wavelengths of visible radiation are significantly stronger than others. The arrangement of these wavelength peaks is respon-

Table 5-6. Electric light source characteristics.

Characteristic	Filament Incandescent	Fluorescent	Compact Fluorescent	High Intensity Discharge
Light Form	Medium-intensity *point* source of light; capable of long-range projection as a directional *cone*	Low-intensity linear source of light; capable of short-range projection of a directional wash	Medium-to-high-intensity linear source of light; capable of medium-range projection of a directional wash	High-intensity *point* source of light; capable of medium-range projection as a directional *cone,* and long-range projection as a general wash; new generation of halide incandescentlike lamps will be capable of long-range projections as directional *cones*
Luminaire Size	Compact to moderately compact depending on particular lamp shape, wattage, and application	Moderately compact to bulky depending on application	Compact to moderately compact depending on application	Moderately compact to bulky depending on application
Auxiliaries	Socket (generally, 1 per lamp; some quartz halogen lamps require 2 sockets)	2 sockets per lamp; at least 1 electromagnetic ballast per 2 lamps (for starting and current control)	1 socket per lamp; at least 1 electromagnetic ballast per 2 lamps	Socket (generally, 1 per lamp; generally, 1 ballast per lamp; 1 quartz lamp and socket for emergency lighting due to slow re-start)
Efficiency (light efficiency)	10–25 lumens/watt	65–92 lumens/watt (includes ballast)	30–79 lumens/watt (includes ballast)	80–110 lumens/watt (includes ballast)
Effective Life	750–2,000 hours	15,000–20,000 hours	10,000–20,000 hours	15,000–20,000 hours

sible for the color of light emitted by the fluorescent lamps and for the color-rendering characteristics of that light. A typical fluorescent lamp's spectrum is illustrated in figure 5-14.

There are two generic classifications of fluorescent lamps: "cold" cathode and "hot" cathode. Cathodes, or electrodes, are the metallic elements at opposite ends of the lamps that produce the arc (electric discharge) within the lamp. Cold cathode lamps require a high voltage to produce an electric arc between the two cathodes. These lamps have generally been used for signs (fig. 5-18), as they are easily bent to form word or object outlines and can be operated in many kinds of environments.

Many times cold cathode lamps are incorrectly referred to as "neon" (neon lamps are filled with neon gas, rather than mercury or other gases, and produce only warm tones of light). Cold cathode fluorescent lighting use has recently increased in architectural applications where architectural details dictate small-diameter, continuous-length linear sources; the visual environment is enhanced by the lamp's clean functionalism; and the festive mood and futuristic look generally associated with the lamp's architectural use is desired. Figures 5-19 and 5-20 illustrate architectural applications of cold cathode fluorescent lighting.

Hot cathode fluorescent lamp types operate on the principle of building up heat on the cathodes before introducing a voltage charge necessary to strike an arc. This permits lower voltage for starting, resulting in more efficient operation. The rapid-start fluorescent lamp is the most common. Their instant-on, good efficiency and long life characteristics combined with a wide variety of lamp color temperatures and color rendering abilities enable the designer to develop attractive, efficient work environments.

Fluorescent lamps, like all electric discharge lamps, require a ballast for operation. The ballast provides the necessary voltage for lamp starting and

Figure 5-18. Electric discharge lamp: cold cathode fluorescent sign.

Figure 5-19. Electric discharge lamp: cold cathode architectural lighting in the floor opening.

Figure 5-20. Electric discharge lamp: cold cathode architectural lighting provides continuous brightness bands around curvilinear shapes.

limits current flow to the lamp, avoiding premature and violent lamp burn-out. With electronic ballasts, there has been a revolution in lighting controls—with the introduction of reliable, energy-efficient dimming and conventional on-off control of each luminaire separately within a space.

Traditionally, designers have neglected the use of the fluorescent sources in what might be considered leisure, lounge, intimate, or residential-type environments. This is, no doubt, because of the many poor color and color rendering characteristics of fluorescent lighting in the past, combined with their traditional high-light level use in the offices of the 1950s, 1960s, and 1970s, when fields of fluorescent lighting became synonymous with work, stress, and strain. The responsible designer following good design principles can successfully incorporate fluorescent sources in many spaces resulting in the appropriate environmental impressions. Characteristics of fluorescent lamps are summarized in table 5-6.

The second family of electric discharge lamps are referred to as high intensity discharge (HID) because of their comparatively high light intensities. HID sources emit only certain specific wavelengths of visible radiant energy and are therefore known as *discontinuous spectrum* sources. The spectral diagram of a metal halide lamp in figure 5-14 illustrates this effect.

The most efficient HID sources used in interior environment applications are of the metal halide family. Mercury vapor lamps and high pressure sodium lamps are rarely used because of the relative inefficiency and poor color of the former and the poor color of the latter. Although used primarily in indirect lighting applications in the 1970s because of their relatively high wattage and high light output, metal halide lamps' uses are increasing in traditional downlight, wall wash, and accent light applications as smaller wattage sources become available. Consequently, direct glare is more tolerable, if not eliminated.

Color Compatibility of Electric and Natural Light

Daylight is quite variable in *whiteness,* as is indicated in figure 5-21. During most of a clear day, however, the mixture of sunlight plus skylight will produce a whiteness range that falls between 4,000°K and 5,500°K (equivalent blackbody temperature). Heavy overcast conditions will produce somewhat higher color temperatures. This range is generally defined as *cool.* In most large working environments where direct lighting is used extensively, matching lamp color temperature with daylight is not necessary. Should the designer attempt to match electric source color with daylight, however, the window transmission effect must be considered (for example, bronze tint windows will exhibit a color temperature lower—warmer—than general daylight).

Color compatibility is primarily important when the designer is dealing with either indirect electric or natural light or both. For example, clerestories tend to light the ceiling, producing indirect natural lighting on occu-

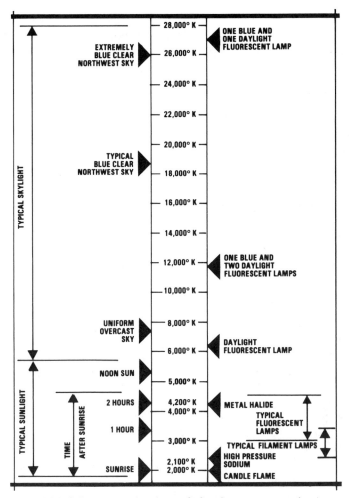

Figure 5-21. Color temperatures (natural-electric source comparison).

pants. If electric lighting of the wall or supplemental electric indirect ceiling lighting are employed, the space will appear more cohesive if the electric sources match the spectral characteristics of the predominant daylight. There may be times when it is desirable to *emphasize* the existence of daylight; therefore, electric sources should not mimic the daylight spectral characteristics.

Material Action

Energy in the electromagnetic spectrum travels at a speed of 186,000 miles per second and moves in a straight line. If this energy is to bend or otherwise change its basic character, this change must be induced through interaction with various materials and forms. This conscious manipulation of energy involves the art and science of optics. This in turn involves the physical manipulation of light by one or both of two methods: reflection and transmission.

Reflection

Reflecting materials can be classified as specular, as in a mirror or polished aluminum; semispecular, as in brushed aluminum; or matte (diffusing), as in blotting paper or matte paint. A reflecting material can also be classified as either high reflectance or low reflectance. This characteristic refers to the percent of incident energy that is reflected in any and all manners from the surface. It should be emphasized, however, that reflectance is quite independent of the specularity or diffusion associated with the surface. Table 5-7 reports the reflectances associated with typical architectural materials.

Specular Reflection

The primary characteristic of specular reflection is the fact that the angle of reflected energy will equal the angle of incident energy. Intensity of reflection will be diminished to the degree that the reflectance is less than 100 percent. With this knowledge, the designer can predict the action of reflected light. Similarly, where specular planes such as polished walls and table tops are involved, the designer can identify the limits of the reflected field of view for each occupant position (see figs. 1-30, 1-31, and 1-35). When specular materials are curved in a convex manner about a compact light source, a focusing action begins to occur for the reflected light. This action can be developed to produce either a narrow, concentrated beam or a broad, diverging beam from a relatively small opening (figs. 5-22 and 5-23). Specular reflectors are therefore typically used for luminaire reflector devices where very precise control is required.

Semispecular Reflection

Semispecular reflection is a combination of specular and diffuse reflection. While some light striking a semispecular surface will be reflected in many directions, a portion of the light will be reflected in a specific direction—following the physical laws discussed under *Specular Reflection*. Semispecular reflectors in luminaires can provide excellent light distribution without the excessive glare and/or lamp-image reflection that may exist with specular reflectors.

Table 5-7. Light reflectance (interior).

	Reflectance (%)
Typical specular materials	
Luminaire reflector materials	
silver	90-97
chromium	63-66
aluminum	
polished	60-70
Alzak polished	75-85
stainless steel	50-60
Building materials	
clear glass or plastic	8-10
stainless steel	50-60
Typical diffusing materials	
Luminaire reflector materials	
white paint	70-90
white porcelain enamel	60-83
Masonry and structural materials	
white plaster	90-92
white terra-cotta	65-80
white porcelain enamel	60-83
limestone	35-60
sandstone	20-40
marble	30-70
gray cement	20-30
granite	20-25
brick	
red	10-20
light buff	40-45
dark buff	35-40
Wood	
light birch	35-50
light oak	25-35
dark oak	10-15
mahogany	6-12
walnut	5-10
Paint	
new white paint	75-90
old white paint	50-70

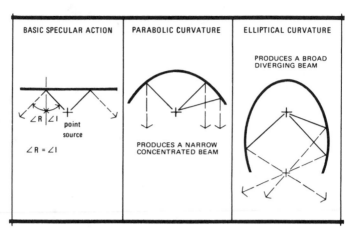

Figure 5-22. Specular reflection.

Figure 5-23. Parabolic reflector: low voltage incandescent filaments surrounded by parabolic reflectors create precise control of light on focal elements.

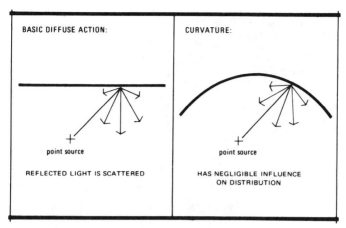

Figure 5-24. Diffuse reflection.

Diffuse Reflection

The primary characteristic of diffuse reflection is the fact that the angle of reflected energy has little relationship to the angle of incidence. The directional integrity of the beam is destroyed, and the energy is scattered (fig. 5-24). Intensity of reflection will again be diminished to the degree that the surface reflectance is less than 100 percent. With this knowledge, the designer knows that a surface of this type will reflect light broadly within a space and that the observer will experience somewhat uniform, integrated surface brightness. For this reason, diffuse materials are typically used for most major portions of spatial surfaces like walls, ceilings, and floors (see fig. 5-25).

When diffusing materials are curved, the curved form serves a function of limiting the distribution of brightness slightly, but there is little or no focusing effect. Regardless of form, then, diffuse reflectors in a luminaire will produce a broad or diverging beam, generally from a relatively large opening.

Transmission

Transmitting materials can be classified as clear transparent; prismatic, as in clear ribbed glass or plastic; or diffusing, as in white glass or plastic. Transmitting material can also be classified as high transmission or low transmission. This refers to the percent of incident energy transmitted through the material.

In a quest for more energy-efficient lighting systems, high transmission materials have become popular luminaire lensing mediums. Although these systems exhibit greater efficiencies they can result in unsatisfactory visual environments. This dissatisfaction generally may be a result of more visible

Figure 5-25. Diffuse reflection: fabric wall surfaces and high reflectance matte ceiling finishes help to uniformly diffuse fluorescent light into this presentation room.

lamp images through the lens and of more pronounced discomfort glare. Visually assessing these luminaires in mock ceilings as well as reviewing VCP data will permit the designer to judge better such high transmission lensed lighting systems.

Prismatic Action

Although rays of energy may pass directly through transparent media such as air, clear water, clear glass, and clear plastic without changing their essential character, the speed of the energy is different in each of these transparent materials. Because of this difference in speed, there is a slight modification in the direction of the energy ray. This speed and subsequent directional change occur at the surface (or surfaces) where the media change takes place. This phenomenon can be observed by placing a straight stick into a clear pool of water. The stick appears to bend at the junction of the water and air.

If a piece of clear glass has parallel faces, therefore, light will bend at the upper surface, while a compensating bend will occur at the lower surface (fig. 5-26, left). As a result, there is a slight displacement of the beam but no permanent change of direction. This type of panel is useful when enclosure is required but no change in beam direction is desired (as is required for spotlight cover plates or for window glass). If a piece of clear glass involves nonparallel faces, however, the light emerges at the lower face with a permanent change in direction (fig. 5-26, center). This prismatic action is a basic principle in lens design.

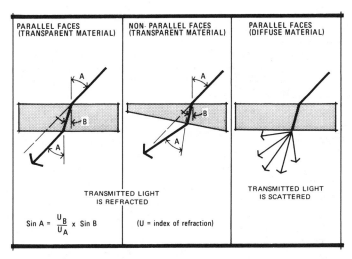

Figure 5-26. Transmission.

Diffuse Transmission

Diffuse transmission like diffuse reflection produces a basic change in an incident ray of light. Regardless of the directional character of the incident ray, the result is a broad scattering of light (see fig. 5-26, right). Also, similar to diffuse reflection, the subjective view of the observer will be that of a uniform, integrated surface brightness. For this reason diffuse materials are typically used for most moderate and large-area transilluminated surfaces. Table 5-8 lists some materials and their respective transmissions and reflectances.

Shielding

Louvers and baffles screen direct light from the eye (fig. 5-27). In this sense they help control direct glare and become a basic tool for manipulating luminaire brightness. Furthermore, since it is generally desirable to limit view of the mechanics of the system, shielding may also be justified from the standpoint of architectural detailing and aesthetics. In general, baffles provide shielding in one direction (that is, along a single viewing axis). Louvers, on the other hand, are a series of baffles or shielding elements arranged in a geometric pattern to provide shielding from many directions. In both cases shielding is effective within a specified zone. This refers to the maximum angle that the eye can be raised above the horizontal without seeing through the shielding system.

Within the shielded zone itself, the brightness of baffle and louver surfaces is determined by the intensity of light reflected from the surface toward the eye. Control of this intensity (to produce high or low system brightness) can

Table 5-8. Light transmission (interior).

	Transmission (%)	*Reflectance (%)*
Clear glass or plastic	80-94	6-10
Transparent colors		
red	8-17	
amber	30-50	
green	10-17	
blue	3-5	
Configurated or prismatic glass	55-90	10-25
Acid-etched glass		
toward source	80-90	5-10
away from source	65-80	10-20
Opal white glass	12-40	40-80
White plastic	30-65	30-70
Marble	5-40	30-70

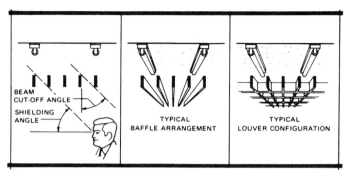

Figure 5-27. Shielding with baffles and louvers.

therefore be manipulated by varying the reflectance of the louver surface (see table 1-5).

Some of the newer baffle and louver materials are simply higher reflectant versions of previous generations. In some cases this means higher-specularity (more mirrorlike) materials are used, resulting in distracting and/or discomfort glare reflected from the baffle or louver. As in the case with new, high transmission lensing materials, this distraction–discomfort glare is not necessarily apparent in test report data, and so visual assessment of a luminaire is desirable. Different baffled or louvered luminaires may exhibit similar high efficiencies but may shield and reflect light in such different ways that one may be totally unacceptable from a glare standpoint, while another may introduce light into a space in a very comfortable, pleasing fashion.

Beam Performance

There is a sense of logic in the coordinated use of materials and energy sources. In this sense, a specific beam of light depends on the integrity of the relationship between reflector forms and finishes, the light source, and the shielding media or facing material. A change in any one of these will alter the characteristics of the beam of light. The material–source relationships most commonly used for interior environmental systems are summarized in figures 5-28, 5-29, and 5-30.

Spatial Influences

In the design and manipulation of the luminous environment, the designer must develop a specification that correlates two major contributing components: the light emitted directly from the source toward the object or surface to be lighted and the light that is incident on the object or surface as a result of reflection from other surfaces in the space.

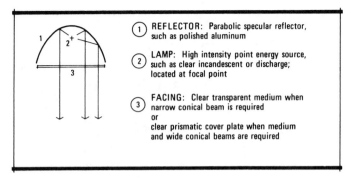

Figure 5-28. Development of a narrow cone of light.

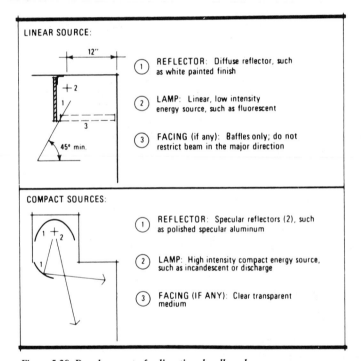

Figure 5-29. Development of a directional wall wash.

Direct Source Emission

Light from a typical point source is emitted in all directions from the source. But most of this light is invisible in the sense that we see only that portion that is emitted toward and which enters the eye. If we assume a complete spatial void, the eye perceives no light except the flux that flows from the source toward the eye; and the remaining light is subjectively lost or

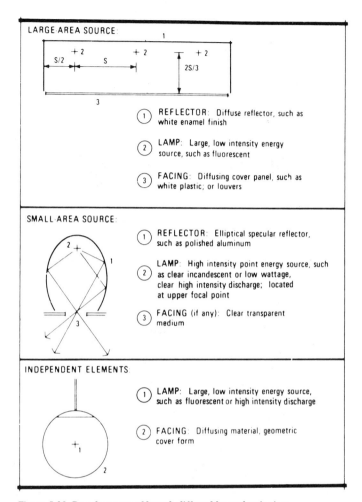

LARGE·AREA SOURCE:

① REFLECTOR: Diffuse reflector, such as white enamel finish

② LAMP: Large, low intensity energy source, such as fluorescent

③ FACING: Diffusing cover panel, such as white plastic; or louvers

SMALL·AREA SOURCE:

① REFLECTOR: Elliptical specular reflector, such as polished aluminum

② LAMP: High intensity point energy source, such as clear incandescent or low wattage, clear high intensity discharge; located at upper focal point

③ FACING (if any): Clear transparent medium

INDEPENDENT ELEMENTS:

① LAMP: Large, low intensity energy source, such as fluorescent or high intensity discharge

② FACING: Diffusing material, geometric cover form

Figure 5-30. Development of broad, diffused beam luminaires.

dissipated in the void. If we introduce a simple diffuse reflecting sphere into the space (fig. 5-31, left), a portion of the previously lost light intersects this form and part of this light is reflected toward the eye. The eye now perceives a partially lighted form; and since the lower portion of the sphere received no direct light, this aspect of the form tends to be imperceptible to the eye.

At this point, the designer can modify the perceived color of the sphere by several methods:

1. Introduce a colorant (a paint or dye) into the finish of the sphere. This colorant serves as a subtractive filter that absorbs certain wavelengths while reflecting others.

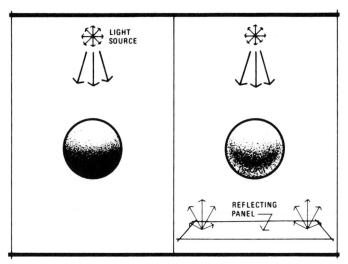

Figure 5-31. Direct and secondary light emission.

2. Introduce a filter (colored glass or plastic) at the light source. This also serves as a subtractive filter, absorbing certain wavelengths and transmitting others toward the sphere.

3. Modify the color additively by introducing additional light sources (lamps, phosphors, and so on). This action shifts the spectral characteristics of the initial source by adding energy at specific wavelengths.

Secondary Sources

If the designer intends for the sphere to be more completely illuminated in order to facilitate perception of the entire form, it will be necessary to introduce a second light source (either primary or secondary) to illuminate the lower portion of the sphere. Figure 5-31, right, suggests the action of a diffuse reflecting plane, which becomes a secondary light source for redirecting light into the shadow area. To some extent, the designer can affect the intensity of the fill light by changing the reflectance characteristics of this surface (between high reflectance white and low reflectance gray). The color of the fill light can be affected by using the reflector surface as a subtractive filter. For example, a red surface will subtract blue and green wavelengths and reflect predominantly red light toward the shadow portion of the sphere.

Manipulation of Visual Space

Similar relationships are developed in the analysis of light in conventional activity-oriented spaces. For example, the illumination of object form by a single direct light source may again be perceived as intense highlight and

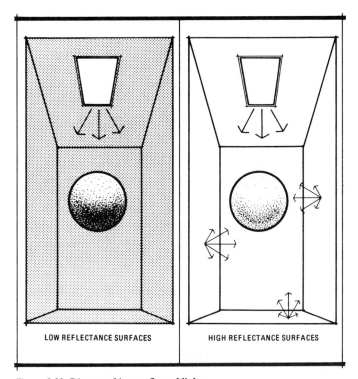

LOW REFLECTANCE SURFACES   HIGH REFLECTANCE SURFACES

Figure 5-32. Direct and interreflected light.

deep shadow voids, while the development of additional primary or second-ary light sources will modify this perception.

The action of diffuse reflecting wall surfaces is, of course, a common influence (fig. 5-32, right). In this sense, the designer can modify the intensity of interreflection (fill light) by changing the reflectance characteristics of various wall, ceiling, and floor surfaces. When all major architectural sur-faces have high reflectance characteristics (near white), the secondary light-ing component will become quite significant and will tend to minimize contrast by filling in the shadows. Recognize that such an environment may use "energy" very efficiently, yet may result in an uninviting, bland space. Table 5-7 lists some typical materials and their respective reflectances. Color of reflected light as well as intensity can be modified by manipulating the room enclosure surface finishes.

Low reflectance room enclosure surfaces (approaching black) will mini-mize interreflection. When narrow beam spotlighting is used in this latter setting, the system will tend to produce highlighted forms within a basic *void* space resulting in a sense of mystery and awe. This selective lighting plan-ning is useful for all types of environments—from dramatic museum settings to high-image offices.

Figures 5-33 and 5-34 are examples of selective light planning in a classroom environment. A uniform arrangement of semi-specular parabolic louver luminaires is positioned and switched to focus primary attention to the work surfaces (fig. 5-33). In figure 5-34, through switching and the use of adjustable accent luminaires along the wall, the horizontal surfaces within the room take a role secondary to that of the vertical surfaces. This setting can be used for more informal discussions, for wall-mounted graphic presentations and, in conjunction with the overhead lighting, to create a sense of spaciousness.

Figures 5-35 and 5-36 are further examples of selective lighting. In figure 5-35, the horizontal plane is lighted, with particular relevant objects-in-space highlighted. The vertical field is deemphasized, resulting in an enclosed, hushed worship environment. As a fundamental contrast in spatial character, figure 5-36 shows the effect of special lighting of high-reflectance vertical surfaces, resulting in a spacious, public-feeling worship environment.

Quantitative Estimates: Nonuniform Lighting Systems

The practice of selectively locating and orienting lighting has increased substantially over the last decade. Increased energy costs are the prime motivator for localized lighting systems. At the same time, such systems tend to establish a visual hierarchy within a space or group of spaces. Illumination in these instances may be relatively high at some areas and low at others, developing a selective light-to-dark contrast between areas and thereby

Figure 5-33. Parabolic louver fluorescent luminaires direct light downward, providing a sense of visual clarity.

creating a visual hierarchy (see chapter 1, sections under *Light-Structure Models* and *Spatial Distribution of Brightness*). Although locally orienting lighting systems at specific task or focal areas may be initiated by intuition, the designer must develop quantifiable evidence to assure that appropriate illuminance (footcandles) is provided and, subsequently, to establish expected luminance ratios.

Figure 5-34. Wall accent luminaires direct light onto the wall surfaces, providing a sense of relaxation.

Figure 5-35. Selective spatial development (subordination of walls).

Figure 5-36. Selective spatial development (emphasis of walls).

Point-by-Point Estimates

Point-by-point analysis at selected locations is the most effective method for quantitatively evaluating spatial lighting effects. In its simplest form, this technique recognizes that the intensity of light will diminish in proportion to the square of the distance from the electric light source to the analysis point and results in a prediction of illuminance at a point due to *only* the *direct* light from a source. This technique can be performed without any sort of calculator or computer.

In its most complex form, the point-by-point technique recognizes not only the intensity of light from the electric light source but will also recognize walls, floor, and ceiling surfaces as "sources" and consider the effects of both reflected light and direct light at a point. This "interreflected-direct light" technique is generally more accurate and generally requires a sophisticated calculator or computer for execution.

For quick, reasonable estimates of illuminance at a point on a horizontal plane and the resultant luminance, due to a *point* source, the following equations apply:

$$E_i = \frac{CP_i \times \cos \theta}{D^2},$$

where E_i is the initial illuminance at the analysis point due to the direct light intensity from a luminaire; CP_i is the initial candlepower intensity (in candelas) of the source in the direction of the analysis point; $\cos \theta$ is the cosine of

Table 5-9. Table of cosines.

Angle of Incidence	Cosine	Angle of Incidence	Cosine
0°	1.000	50°	0.643
5°	0.996	55°	0.574
10°	0.985	60°	0.500
15°	0.966	65°	0.423
20°	0.940	70°	0.342
25°	0.906	75°	0.259
30°	0.866	80°	0.174
35°	0.819	85°	0.087
40°	0.766	90°	0.000
45°	0.707		

the angle of incidence (the angular displacement from perpendicular incidence) (see table 5-9); and D is the distance from the light source to the analysis point (in feet); and

$$L_i = E_i \times R,$$

where L_i is the approximate initial luminance (in footlamberts; assumed to be a completely diffuse reflecting surface—this equation does not apply to specular materials like chrome, aluminum, or glass); E_i is the initial illuminance (in footcandles); and R is the reflectance (in decimal form, $0.8 = 80$ percent reflectance) of the analysis surface.

In most cases, more than one light source will contribute to the illuminance at a point. For these situations, the calculation must be performed for each source in the vicinity of a point, and then the illuminances are added together for the total illuminance at a point due to several point-source luminaires.

Illuminance at a point can be estimated for area and linear sources, although this requires discretization of the source into small sources that approximate point sources. For example, a 12- by 48-inch ceiling-mounted fluorescent luminaire could be discretized into four equal sections, each measuring 12 by 12 inches. By dividing the candlepower values by four and assigning them to each 12- by 12-inch segment, four different point calculations could be performed to find the illuminance at a point. As more luminaires are considered and as wall, floor, and ceiling reflected light is taken into account, more rigorous calculations are required—resulting in the need for calculator or computer programs.

Quantitative Estimates: Uniform Lighting Systems

Uniform lighting systems were popular in the 1950s and 1960s when energy was inexpensive and when very high light levels were thought necessary for productivity and health. Today, uniform lighting systems offer acceptable solutions when flexibility is necessary and overhead ceiling-mounted lighting systems are convenient. Since these systems usually provide roughly equal illuminance over the floor or workplane area, the lighting designer and interior designer must very carefully plan focal centers and visual hierarchy throughout such spaces to avoid bland, uninteresting, and informationless visual environments.

Maximizing light usage of uniform lighting systems requires medium to high surface reflectances. Table 5-10 indicates typically recommended reflectances for work environments, regardless of the type of lighting system. These reflectances in combination with efficient uniformly arranged luminaires will result in a reasonably energy-efficient, satisfactory working visual environment.

A calculation technique known as *Lumen Method* or *Zonal Cavity Method* exists that can account for the effects of light reflection from walls, floor, and ceiling, effects of dirt build-up on the lighting equipment, effects of lamp aging, and effects of luminaire efficiency, resulting in an *average, maintained* illuminance value. Note that in the actual environment it is unlikely that light level readings will match the *average*—only if illuminance is measured at a number of points and then averaged will a value close to the predicted (calculated) average be obtained. This calculation procedure is only valid when luminaires are uniformly distributed throughout a space.

Operational Characteristics

The quantity of light produced for a given input wattage varies with the type of lamp used (see table 5-6). Fluorescent and high intensity discharge sources produce light from three to five times more efficiently than incan-

Table 5-10. Recommended reflectances for general systems.

	Reflectance Range (%)
Ceiling finishes	70-90
Wall finishes	40-70
Floor finishes	10-40
Desk and bench tops	20-40

NOTE: Indicated reflectance ranges refer to matte (diffuse) surfaces and finishes.

descent sources. For this reason, fluorescent and HID sources are often more economical for general lighting systems, particularly in air-conditioned (cooled) space. Since general (uniform) lighting systems may lack the visual interest and stimulation associated with nonuniform systems, the designer will need to introduce supplementary accent lighting systems or accent finishes to provide the desirable sparkle, highlight, and shadow. Correct material, form, and color selection can contribute to a visually stimulating environment.

Quantitative Performance

Quantitatively, illumination is usually discussed in terms of footcandles at the worksurface. Footcandles are the quantity of lumens per square foot of surface. For many commercial and industrial activities, this surface is a horizontal desk or plane approximately 30 to 36 inches above the floor (refer to table 1-8 for suggested illuminance levels for various activities). When estimating lighting systems' performances in given spaces, the designer needs to account for losses due to luminaire limitations, surface reflectances, and room proportions.

Losses within the Lighting Equipment

Light is partially obstructed and absorbed by the luminaire itself (by lenses, louvers, reflectors). As a result, the total lumens emitted from the luminaire are less than the lumens generated by the lamp.

Losses at Room Surfaces

Although some of the light from the luminaire directly impinges on the work surface, some amount may be reflected at least once from the ceiling, walls, and floor. Since no material reflects all of the light that strikes it, each reflection will involve some loss of light through absorption. Therefore, high reflectance finishes reduce these losses and improve system efficiency. Furthermore, light may be reflected from surface to surface before striking the workplane. This interreflection is better sustained with high reflectance finishes. The designer is warned that although the use of specular (directionally reflective) materials may be desirable from an "image" standpoint or for increasing light striking the workplane, their use must be carefully planned to avoid offensive glare problems.

Losses Due to Room Proportions

The greater the floor area for a given ceiling height, the more efficient will be the uniform light system in delivering light to a horizontal worksurface. In a large, low-ceilinged room, a substantial portion of the light from the luminaires directly reaches the workplane without reflection (and without subsequent reflection losses) from room surfaces. On the other hand, in a high-ceilinged, narrow room, no-furniture-partitions space, a higher proportion of the light strikes the walls and other spatial surfaces. This condition results in reduced lighting efficiency.

Losses Caused by Light Distribution

To some extent the distribution characteristics of the lighting equipment determine (in part) the amount of the initially emitted light that will be reflected from various room surfaces. If the luminaire has a very narrow light distribution pattern, very little light will strike the wall surfaces, resulting in minimal light loss. Recognize, however, that although such systems may statistically produce a higher average footcandle value within a space than wide-distribution or indirect distribution lighting systems, this does not indicate the systems' abilities to meet other performance criteria, including visibility, visual comfort, luminance ratios, and psychological requirements. In fact, such narrow-distribution systems can lead to a darkened room appearance, since little or no light falls on the walls. With these systems it may be desirable to add wall lighting.

The Coefficient of Utilization

The effects of the previously discussed losses can be summarized in a single design factor, the *coefficient of utilization*. This factor is the percentage of generated lamp lumens that will actually reach the horizontal workplane for a given lighting system, room proportions, and surface reflectances. A coefficient of utilization (CU) of 0.65 means that 65 percent of the lamp lumens are reaching the workplane.

The formula used for estimating or predicting the illuminance at a horizontal workplane created by uniform lighting systems is:

$$E_i = \frac{LL \times CU}{A}$$

where E_i is the average initial illuminance (in lumens per square foot, or footcandles); LL is the initial total lamp lumens produced by all lamps in the lighting system (provided in the lamp manufacturers' literature); and CU is the coefficient of utilization (provided in the luminaire manufacturers' literature).

As indicated in figure 5-37, coefficients of utilization depend on room surface reflectances and on room proportion.

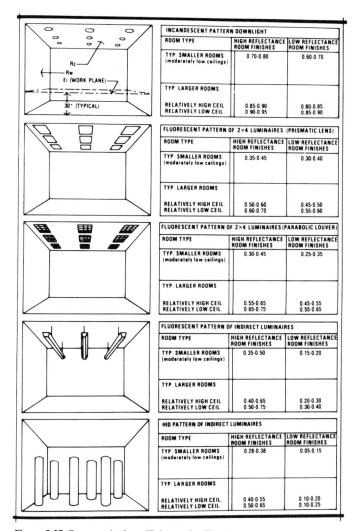

INCANDESCENT PATTERN DOWNLIGHT

ROOM TYPE	HIGH REFLECTANCE ROOM FINISHES	LOW REFLECTANCE ROOM FINISHES
TYP. SMALLER ROOMS (moderately low ceilings)	0.70-0.80	0.60-0.70
TYP. LARGER ROOMS		
RELATIVELY HIGH CEIL. RELATIVELY LOW CEIL.	0.85-0.90 0.90-0.95	0.80-0.85 0.85-0.90

FLUORESCENT PATTERN OF 2×4 LUMINAIRES (PRISMATIC LENS)

ROOM TYPE	HIGH REFLECTANCE ROOM FINISHES	LOW REFLECTANCE ROOM FINISHES
TYP. SMALLER ROOMS (moderately low ceilings)	0.35-0.45	0.30-0.40
TYP. LARGER ROOMS		
RELATIVELY HIGH CEIL. RELATIVELY LOW CEIL.	0.50-0.60 0.60-0.70	0.45-0.50 0.55-0.60

FLUORESCENT PATTERN OF 2×4 LUMINAIRES (PARABOLIC LOUVER)

ROOM TYPE	HIGH REFLECTANCE ROOM FINISHES	LOW REFLECTANCE ROOM FINISHES
TYP. SMALLER ROOMS (moderately low ceilings)	0.30-0.45	0.25-0.35
TYP. LARGER ROOMS		
RELATIVELY HIGH CEIL. RELATIVELY LOW CEIL.	0.55-0.65 0.65-0.75	0.45-0.55 0.55-0.65

FLUORESCENT PATTERN OF INDIRECT LUMINAIRES

ROOM TYPE	HIGH REFLECTANCE ROOM FINISHES	LOW REFLECTANCE ROOM FINISHES
TYP. SMALLER ROOMS (moderately low ceilings)	0.35-0.50	0.15-0.20
TYP. LARGER ROOMS		
RELATIVELY HIGH CEIL. RELATIVELY LOW CEIL.	0.40-0.65 0.50-0.75	0.20-0.30 0.30-0.40

HID PATTERN OF INDIRECT LUMINAIRES

ROOM TYPE	HIGH REFLECTANCE ROOM FINISHES	LOW REFLECTANCE ROOM FINISHES
TYP. SMALLER ROOMS (moderately low ceilings)	0.28-0.38	0.05-0.15
TYP. LARGER ROOMS		
RELATIVELY HIGH CEIL. RELATIVELY LOW CEIL.	0.40-0.55 0.50-0.65	0.10-0.20 0.10-0.25

Figure 5-37. Some typical coefficients of utilization.

Maintenance of Light Output

Since initial illuminance estimates are based on the lumen output of new lamps in clean luminaires and in clean spaces, these values are only a prediction of what can be expected through the first several months of operation. Depending on the cleanliness of the environment and the cleaning and lamp-replacement schedule (as well as the type of lamp and luminaire), a design allowance is usually made to compensate for the expected reduction in light output. An assumed reduction to between 75 and 85 percent of initial light is typical for clean or air-conditioned environments with fluorescent luminaires. For more adverse conditions and for most HID lamps, an assumed reduction to between 55 and 75 percent may be more appropriate, particularly if the designer is concerned with ballast and thermal effects in luminaires. These design reduction values are referred to as maintenance factors.

Using this factor then, the formula for predicting average illuminance produced by a system after a period of operation is:

$$E_m = E_i \times LLF,$$

where E_m is the maintained average illuminance; E_i is the initial average illuminance; and LLF is the light loss or maintenance factor.

Losses Caused by Objects in the Room

Introducing objects in a room will generally cause a reduction in room lighting levels and brightnesses. This can be a significant problem in large open office-plan areas with a lot of relatively high movable partitions. Of course, the human body will help to reduce task lighting levels, by blocking light from behind. These losses can approach 25 percent, while typically they account for a 10-15 percent reduction.

Illuminance Criteria Compliance

With the use of calculators and computers there exists a temptation to base design on objective illuminance criteria to meet the desire for pinpoint accuracy. The designer must recognize that estimates or predictions are no more than that. Calculating quantities like illuminance to even a single decimal point is generally meaningless. Systems that may miss meeting illuminance criteria by 5 or even 10 percent (upon calculation) should not be discarded from consideration on this point alone. Compliance with other criteria should be investigated and, if necessary, a selection matrix developed to list criteria by priority of importance.

When checking final installations against calculations, do not be alarmed if the measured illuminance is up to 25 percent higher or lower than calculated. Recognize that the calculation procedure discussed above determines *average* illuminance. Also recognize that metering devices themselves may be as much as 5 to 10 percent in error. Further, calculation techniques can be as much as 10 to 20 percent in error.

Estimates of Spatial Brightness and Interreflection

The coefficient of utilization refers only to the effect of uniform lighting systems in providing light on a single *horizontal* plane. The CU factor

Table 5-11. Wall and ceiling factors.

	Wall Factors	*Ceiling Factors*
Downlight pattern	0.30-0.45	0.12-0.20
Pattern of spheres	0.85-1.00	0.80-1.20
Fluorescent pattern	0.30-0.50	0.12-0.25
Luminous ceiling	0.60-0.75	—
Uplight	0.60-0.75	1.50-3.00[1]
		2.50-6.00[2]

1. For high-reflectance room finishes.
2. For low-reflectance room finishes.

therefore constitutes one index of system performance relative to typical seeing tasks but makes no allowance for the spatial factors that also affect the visual environment (see chapter 1).

To assist further evaluation, the factors shown in table 5-11 can be used to develop general approximations of wall and ceiling luminance for the representative general systems noted in figure 5-37. Using these factors, the formula for estimating the effect of general lighting systems in producing wall luminance is:

$$L_w = E_{wp} \times WF \times R_w,$$

where L_w is the approximate average wall luminance at the midpoint between floor and ceiling (in footlamberts); E_{wp} is the average horizontal illuminance at the workplane (in footcandles); WF is the approximate wall luminance factor (see table 5-11); and R_w is the reflectance of the wall for which the luminance is being calculated (percent).

Similarly, the formula for estimating ceiling luminance is:

$$L_c = E_{wp} \times CF \times R_c,$$

where L_c is the approximate average ceiling luminance (in footlamberts); CF is the approximate ceiling luminance factor (see table 5-11); and R_c is the reflectance of the ceiling (percent).

Lighting System Design

Designing a lighting system, as previously discussed, involves much more than performing some prediction calculations. The lighting system must integrate structurally with the building and aesthetically with the environment. The lighting system must provide for the needs of the space occupants—in both a physiological sense and a psychological sense.

Intuitive Factors Relating to Lighting System Design

In addition to functional aspects related to the light produced in the space, luminaires, their layout, and other physical components of the lighting system can be potentially prominent factors in the visual composition of the space. The physical elements of a lighting system must therefore be analyzed by architectural-interior design standards as well as for their engineering function and performance (see chapter 1 for discussion of *Spatial Order and Form*). In this regard, a study of architectural history reveals two basic approaches to lighting system architectural integration: the *visually subordinate system,* and the *visually prominent system.*

Visually Subordinate Lighting Systems

Some design concepts reflect an effort to introduce light so that the occupant will be conscious of the *effect* of the light but not the *source.* For example, in some Byzantine churches small, unobtrusive windows were placed at the base of a dome to light this large structural element. The brilliant dome then became a dominant visual factor in the space. Serving as a huge reflector, the dome (not the windows) became the apparent primary light source for the interior space. Similarly, in some Baroque interiors the observer's attention was focused on a brightly lighted decorative wall, while the window on an adjacent wall that provided the illumination was somewhat concealed from normal view. In both cases the objective was to place emphasis on the surfaces to be lighted while minimizing any distracting influence of the light itself.

In these historical instances, daylighting systems were involved. The same design attitude can be seen, however, in the development of some electric lighting systems (see, for example, figs. 5-25 and 5-38). These systems use compact, directional sources, hidden sources or luminaires with low-brightness shielding devices. Such equipment directs light toward a specific surface, plane, or object, emphasizing these areas with little distracting influence from the lighting equipment itself. Inherently, then, the space is visually defined as a composition of *reflected* light pattern (horizontal and vertical).

Accurate Delineation of Spatial Form and Surface Character

The *form* of the light distribution pattern (such as a *cone* or *area wash* of light) should relate logically to the form of the affected surface. A wall wash similar to that shown in figures 5-20 and 5-39 best enhances the architectural

Figure 5-38. The visually subordinant lighting system.

Figure 5-39. Delineation of spatial form.

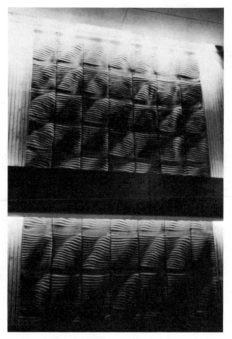

*Figure 5-40. Incandescent lamps graze this tex-
tured surface, delineating the surface character.*

form while providing sufficient spatial definition and illuminance for circulation through the space. If not properly planned, *scallops* and similar light
patterns can tend to introduce visual confusion regarding spatial form and
definition. Of course, special effects including sparkle and highlight may be
required to instill a sense of vitality, and so some light patterning is expected.
Generally, the observer expects surfaces to be integrated, complete forms—
not pierced by meaningless light patterns.

The lighting system should also perform in sympathy with the surface
character involved, with grazing light used to complement a textured surface
(fig. 5-40), and a more diffuse or frontal light used for a flat surface.

Visually Prominent Lighting Systems

A light source or luminaire may attract attention to itself, even to the
extent that such elements become particularly dominant factors in the
design, as do the sconces shown in figures 5-16 and 5-41. Such systems must
be carefully integrated into the interior environment so as to eliminate visual
clutter (as discussed in chapter 1). The lighting system itself is an architectural feature in this case, adding to visual stimulation and, hopefully, to

Figure 5-41. Visually prominent lighting system helps attract observer attention to the other end of the room.

spatial organization. These visually prominent systems must be carefully integrated both visually and physically with the architecture. Otherwise, the lighting equipment appears to be an after-thought to the architectural design.

Visual prominence should not be achieved with high brightness sources. The glare from such equipment minimizes the overall effectiveness of the environment and acts to force people away from the space.

REFERENCES

Kaufman V. E., ed. 1987. *Illuminating Engineering Society Handbook: 1987 Reference Volume.* New York: IES.

Steffy, G. R. 1990. *Architectural Lighting Design.* New York: Van Nostrand Reinhold.

CHAPTER 6

Sound Generation and Control

Sound is initially generated by a vibrating source, such as the action of the vocal chords or the action of a musical instrument. These vibrations induce small atmospheric changes that alternately vary above and below normal atmospheric pressure[1] (fig. 6-1). The average deviation in pressure is called *sound pressure* (intensity); and, by causing the listener's ear drum to vibrate in sympathy, these pressure variations produce a sensation of hearing.

The tone of the signal is determined by the rate at which the pressure alternates above and below the ambient atmospheric condition. This variable is called *frequency* and is determined by the rate of vibration at the source. High rates of source vibration will produce short wavelength, high-pitched sounds; low vibration rates will produce a low-pitched or low frequency sound.

Sounds may originate within the immediate space—examples are conversational speech and other human or mechanical sounds associated with an activity. Sounds that originate externally include vehicular traffic noise, railroad or aircraft noise, sounds associated with playground activity, wind and other natural sounds, and noises associated with activities in adjacent or nearby rooms.

In sonic design, the designer must develop a spatial specification that assists in defining the meaningful sound signals required for communication and orientation, insures that these meaningful signals are perceived against an acceptable sonic background, and minimizes irrelevant and disruptive sounds (noise).

---

1. Sound vibrations can also be transmitted through liquid and solid mediums. Some such transmitting medium must be present, however, as can be demonstrated by operating a bell inside of an exhaustible jar. As long as air remains in the jar, the bell is heard ringing. But when the air is withdrawn, the bell continues to operate without being heard.

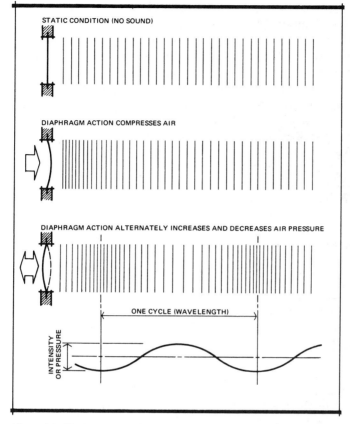

Figure 6-1. Generation of sound.

REINFORCING THE PRIMARY COMMUNICATION SIGNAL

Initially, sound will travel outward in all directions from the source in a manner somewhat comparable with the effect of water waves as they travel outward from the point at which a stone is dropped. As the wave travels, pressure (and therefore intensity) diminishes with distance.

But when the sound originates within an enclosure, the initial wave pattern may also be intersected by other waves that are reflected from the enclosing shell. At each of the enclosing surfaces, sound energy will be partially absorbed and partially reflected. The overall spatial effect of this action will significantly influence the sonic quality of the space because this determines the tendency of sound signals to be either subtractively dampened or to mix additively and reinforce (or muddle) the direct signal.

For a sustained sound in an enclosed space, then, the ear hears both *direct sound* (as diminished by distance) and *reflected sound* (as redirected and diminished by the action of room surfaces). When the source and listener are in close physical proximity, the direct component is dominant and may, in itself, be totally adequate for effective communication. But as the distances become greater, it becomes increasingly necessary for the designer to develop a method to implement or reinforce weak signals.

This reinforcement can be produced by one or both of two methods: by *natural amplification* techniques that utilize surface forms and materials selectively to reflect peripheral sound energy in a manner that will reinforce the direct signal, and by the use of various *electronic amplification* techniques that either increase the intensity of the initial source or transmit the signal to more remote locations.

Material Action

Reinforcement or suppression of sound depends on careful selection and placement of sound reflectors and barriers. When a sound wave strikes a sizable surface or object, part of the energy is absorbed and the remainder is transmitted or reflected. Signal distribution within an enclosure can therefore be partially manipulated through reflection; while intensity and frequency balance can be partially manipulated through absorption.

Reflection

Sound, like light, is reflected approximately according to the conventional laws of reflection (that is, the angle of incidence equals angle of reflection). If the reflecting surface is a plane, the plotting of apparent *sound images* behind the plane is a useful technique for evaluating the characteristics of that surface in redirecting sound (fig. 6-2).

Surface Configuration

The form or configuration of room surfaces will therefore affect the sound-reinforcing properties of the enclosure—redirecting and focusing energy, or diffusing and mixing sound.

Hard, nonporous, and *continuous* surfaces (such as large panels of plaster, wood, or concrete) are highly reflective over the entire range of relevant frequencies. Such surfaces are somewhat analogous with a *specular* reflecting surface in lighting design.

Hard, nonporous but *irregular* surfaces (such as coffered plaster ceilings, or thin-shell concrete folded plates) are, on the other hand, analogous with a white *diffusing* material in lighting (fig. 6-3). In this case, effective diffusion

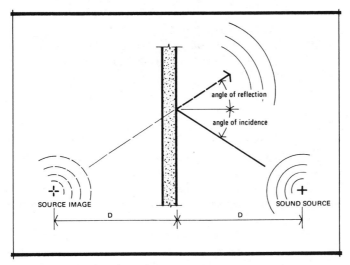

Figure 6-2. Reflection of sound.

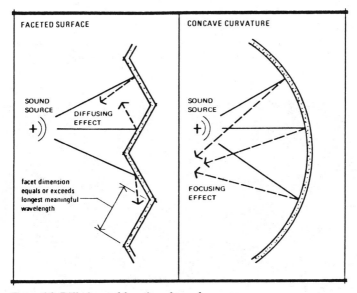

Figure 6-3. Diffusion and focusing of sound.

of sound will result from large-scale surface irregularities of normally at least 3–4 feet or more across and 6 inches or more in depth. The dimension of each facet or reflecting plane must equal or exceed the longest relevant wavelength; so small-scale configurations will only affect high frequency (short wavelength) sounds.

Concave Surface Curvature

Continuing the general analogy with the action of light, large unbroken concave surfaces will tend to redistribute sound energy in an irregular manner. As a result, these surfaces do not satisfy the normal requirements of uniform signal distribution.

For this reason, focusing shapes should be avoided. Or, if this is not possible, superimposed reflecting panels or large-scale surface modulation may be required (fig. 6-4). Still another method is the development of curved forms as *acoustically transparent* surfaces, such as occurs with the use of separated wood strips backed by an air space that includes appropriate angled reflective panels or highly absorbent materials.

Natural Sound Reinforcement

When sound is emitted from a single source toward an audience, the sound energy received by those in the rear is reduced both by normal decay

Figure 6-4. Large-scale diffuser panels in a concave interior.

due to distance and by the absorption effect of the audience in front. The Greeks and Romans minimized these losses by placing the audience on a steep slope in an outdoor theater. This action minimizes the effect of audience absorption and makes signal intensity almost directly dependent on the distance from the source. A similar result can be achieved by placing the sound source high above the audience, but this may cause an uncomfortable or inappropriate visual relationship.

This latter effect is simulated in interior spaces when a hard reflective ceiling is utilized as a *sound mirror* (fig. 6-5). When high intensity signal distribution is required, this mirror can be shaped and refined in a number of ways in order to maximize the area of reflective surface that individual listeners will *see* from their listening positions in the audience.

When differing sound signals are emitted concurrently by a large group, such as an orchestra or chorus, the sound-reflective surface serves to diffuse

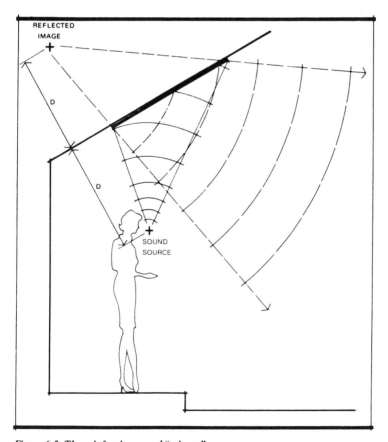

Figure 6-5. The reinforcing sound "mirror."

and mix the sound. Diffusion is important in providing consistent communication with the audience—to insure that various listeners are hearing the same thing. Diffusion is also important for the performers themselves, for it is essential that all performers hear each other. This latter need generally leads to the use of overhead or vertical faceted reflectors in the immediate vicinity of the performing group. This may take the form of coffers or folded plates.

Essentially, then, design for natural amplification in an enclosed space leads toward the use of hard and reflective surfaces that redirect sound from the source to the listener and provide moderate diffusion to insure adequate mixing and uniformity over the entire audience.

Ray Diagrams

Reflected sound in an enclosed space can be studied through *ray diagramming* (fig. 6-6), a method of analysis that recognizes and utilizes the previously discussed sound reflecting characteristics of major room surfaces. As noted earlier, the smallest dimension of an individual reflector facet or panel must be greater than the wavelength of the lowest significant frequency. The following formula is useful for determining this wavelength.

$$\text{Wavelength (feet)} = \frac{\text{Velocity of sound}}{\text{Frequency}} = \frac{1{,}140 \text{ ft/sec}}{\text{(cycles per second)}}$$

This minimum dimension requirement generally leads to panel dimensions of 4 feet or more in a space that is intended to facilitate improved verbal communication. Music may require slightly larger panels because lower frequency sounds are significant. Reflector panels or facets must be organized within a spatial *volume* that is determined by *reverberation time* requirements (discussed later in this chapter).

Special objectives to be included in the design of a natural amplification system are the following:

1. To develop ceiling and wall forms to maximize the useful reflector surface.
2. To avoid *standing waves,* where sound is reflected back and forth between two parallel surfaces of approximately equal dimensions
3. To avoid focusing effects, such as those caused by concave curvature
4. To provide for blending of sound and diffusion through faceting of major surfaces (faceting provides overlapping of reflected sound images)
5. To arrange useful facets so that first reflection sounds arrive 0.035 seconds or less after the direct signal

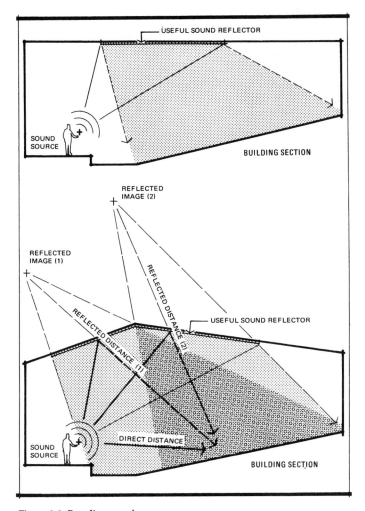

Figure 6-6. Ray diagramming.

6. To manipulate remote materials and forms in a manner that will prevent echo or *muddling* conditions

7. To manipulate materials and space volume to provide the proper reverberation time

See also figure 6-6.

When projection of speech is important, it is desirable to arrange the reflecting surfaces so that the maximum first reflection sounds arrive at each seating location within 0.035 seconds after the direct signal. This means that the reflected path should generally not exceed the direct path by more than 40 feet. If this dimensional limit is exceeded, the reflected signals may tend to become a muddling or blurring influence at the listening location in question. Furthermore, when the reflected path is more than 65 or 70 feet longer than the direct path, a distinct echo will be possible. In both of these latter situations, the offending surface or surfaces should be treated with absorbing materials or otherwise angled or configurated to *ground* the conflicting sound reflections.

Estimating the Perception of Loudness

The decibel scale is logarithmic as a measure of power level. A 0-dB sound at the base frequency of 1,000 cycles per second is barely audible; a 10-dB sound is ten times as intense; a 20-dB sound is ten times the intensity of the 10-dB sound (100 times the intensity of a 0-dB sound); 30-dB sound is ten times the intensity of a 20-dB sound; and so on.

This scale also provides a method for estimating the *subjective* effect of sound contrast. A listener interprets a 10-dB increase as approximately twice as loud; a 20-dB increase is perceived as approximately four times as loud, and so on.

As a method for interpretation of both power level and subjective interpretation, this scale is particularly useful for quick estimates (table 6-1). The following example will serve to illustrate. Since 6 dB approximates a fourfold change in power level, and intensity tends to diminish with the square of the distance, the designer can anticipate a drop of approximately 6 dB for every doubling of the distance from the source to the receiver. The subjective sense of change or contrast for a given listener location can then be interpreted from the *apparent change* column of table 6-1.

Additive Sonic Signals

In ray diagramming, the direct signal (diminished by distance) and the reflected signals (diminished by distance and surface absorption) are approximately additive if the arrival times are less than 0.035 second apart. However, since decibels are logarithmic units, they cannot be added directly. Figure 6-7 is provided to assist in estimating additive intensity.

Electronic Amplification

As a general rule, seating capacities of 600 or less should be capable of adequate signal reinforcement by natural means, while occupant capacities of 1,000 or more will nearly always require some electronic amplification. Electronic systems will probably also be required for sound projection in large, low-ceiling, and *soft* rooms.

Table 6-1. Subjective perception of changes in loudness.

Change in Measured Intensity (decibels)	Apparent Change (subjective sense of change)	Actual Change in Power Level (times base power level)
1	Base unit of measurement	1.25 (or 0.8)
3	Barely perceptible change	2 (or 0.5)
6	Change perceptible with doubling (or halving) the distance from the source	4 (or 0.25)
7	Clearly perceptible change	5 (or 0.2)
10	Change perceptible as *twice as loud* (or *half as loud*)	10 (or 0.1)
20	Change perceptible as *four times as loud* (or *one-fourth as loud*)	100 (or 0.01)
30	Change perceptible as *eight times as loud* (or *one-eighth as loud*)	1000 (or 0.001)
40	Change perceptible as *sixteen times as loud* (or *one-sixteenth as loud*)	10^4 (or 10^{-4})
100	Change perceptible as *1,000 times as loud* (or *0.001 times as loud*)	10^{10} (or 10^{-10})

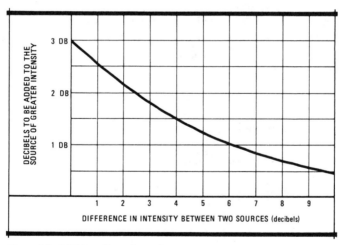

Figure 6-7. Additive effect of sound sources.

Generally, a single *central* speaker (or speaker cluster) located near the visual source is preferable over a distributed speaker system. This central approach will produce an amplified sound with the same basic directional and time characteristics as the original emitted sound. All listeners should have "line-of-sight" to this central speaker (or cluster), because higher frequency amplification is very directional.

To prevent *feedback*, which is caused by amplified sound being picked up by the microphone, large highly directional speakers should be used (typical size: $8 \times 8 \times 3$ feet). This approach is sometimes described as a *high level system*.

In low ceilinged rooms, however, a central speaker system may not be practical. A distributed system of small individual speakers (typically 8 inches) should be mounted so that they project sound directly down from the ceiling. This approach is sometimes described as a *low level system*.

In either case, care should be exercised to prevent a late arrival of *live* direct sound. This lag behind amplified sound can produce muddling or echoes. Correction can be provided either by minimizing the intensity of the live sound or by introducing a slight time lag in the amplification. Electronic sound reinforcing systems should not depend on room surface reflections. "Dead" or soft room conditions should be provided.

Generally, a listener without visual contact with the source will associate the location of the sound source with the first sound that reaches him, even if subsequent arrivals are more intense. This precedence effect must be considered in adjusting time lags in the amplification system.

CONTROL OF INTERNALLY GENERATED NOISE

Sonic communication is facilitated by a high signal-to-background ratio. In this sense, natural or electronic amplification is useful for reinforcing the primary signal itself. But in many cases, improved control of background noise is also logical and effective as a supplementary (or primary) spatial treatment.

Material Action

Irrelevant signals or noise can be partially suppressed through the subtractive action of surface absorption. This action produces a decrease in the pressure of a reflected sound relative to the pressure or intensity of the incident signal.

Speed of energy travel is a constant and is not affected by absorption.

Absorption

Changes in the directional characteristics of a sound wave do not reduce its intensity level, except to the extent that some of the energy is absorbed by the surfaces it strikes. The proportion of incident sound pressure that is absorbed during a given reflection is called the *sound absorption coefficient* of that particular reflecting surface. For example, hard, nonporous surfaces such as plaster, glass, wood, concrete, and most sheet plastics generally have low absorption coefficients of 0.05 or less (that is, 95 percent or more of the incident energy is reflected or transmitted).

When the designer needs a material to reduce or dampen the internal noise level, he is looking not for a hard, nonporous material, but for a porous or soft material that will permit sound waves to penetrate. This penetration causes sound to lose energy by frictional drag before it reenters the room. Much higher absorption coefficients are involved. Acoustical blankets and acoustic tile are examples of materials that exhibit a high absorption coefficient, as are carpeting, heavy fabrics, clothing, and upholstery. The plane of application (vertical versus horizontal) influences the effectiveness of absorption. Figure 6-8 illustrates a typical arrangement of suspended, vertical absorbing baffles.

Panel flexure by thin plywood or plastic sheeting may also produce some dampening of low frequency sounds. But this characteristic can be utilized for noise reduction only when the flexing panel is backed by a relatively large unoccupied space (such as a lighting cavity). This is a somewhat unique case, however, and noise reduction is more commonly produced *within* the absorbing material than by surface flexure.

Characteristics of Absorbing Materials

The absorption qualities of porous materials are basically determined by material thickness and surface finish (fig. 6-9).

• *Backing:* When porous materials are mounted directly on a hard surface (such as concrete, plaster, or gypsum board), the system produces good absorption of middle and high frequencies. However, low frequencies are absorbed very little.

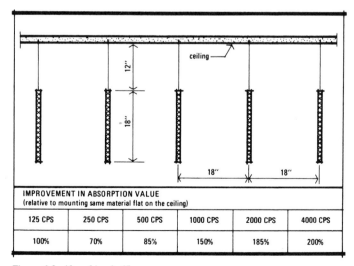

IMPROVEMENT IN ABSORPTION VALUE (relative to mounting same material flat on the ceiling)					
125 CPS	250 CPS	500 CPS	1000 CPS	2000 CPS	4000 CPS
100%	70%	85%	150%	185%	200%

Figure 6-8. Absorbing baffles.

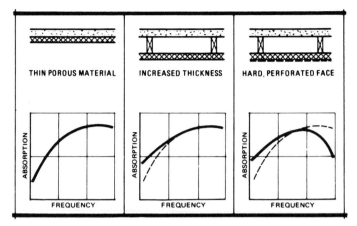

Figure 6-9. Installation and finish variables.

• *Thickness:* Low frequency absorption is improved when the thickness of the absorbing material is increased. This can be done either by providing a thicker porous blanket or by providing an air space behind the material.

• *Facing:* A hard, perforated facing or finish will reduce absorption of high frequency sounds. This varies somewhat with the size and spacing of openings (holes) in the facing.

Very thin plastic, paint, or paper films can be used as protective coverings for maintenance purposes. In these cases, relatively little energy is reflected and most of the incident sound is transmitted directly to the porous core.

Room Absorption

The term *room absorption* is an expression of the total absorption influences within a room. To determine total absorption, the area of each room surface is multiplied by its absorption coefficient (for example, 100 sq ft $\times$ 0.60 = a surface absorption of 60 sabins).[2] The sum of these surface effects is added to the absorption of significant individual objects and furnishings in order to derive a total *room absorption*. In large spaces, there may be a slight addition due to the effect of the air mass itself.

This relationship is summarized in the following formula:

$$\Sigma SA = S_1 A_1 + S_2 A_2 + S_3 A_3 + \ldots,$$

2. One sabin is the absorption value of one sq ft of surface that has theoretically perfect absorption qualities (that is, an absorption coefficient of 1.0). Therefore, the above 100 sq ft-example, which provides an absorption of 60 sabins, is the equivalent of a 60-sq-ft surface that has an idealized absorption coefficient of 1.0.

where ΣSA is the total room absorption (in sabins); S_1, S_2, S_3, etc., is the area of each surface or object; and A_1, A_2, A_3, etc., is the absorption coefficient of each surface or object (table 6-2).

In critical situations, such calculations should be developed for each frequency band. However, as a general indication of room performance, the 500 cps band is usually evaluated as typical.

The appropriateness of various room absorption totals can be interpreted with the assistance of table 2-1.

Estimating Noise Control Value (Internal Noise)

Multiple room reflections tend to increase the general noise level in active spaces and obscure the clarity, location, and identity of meaningful signals. For most *general purpose spaces,* then, it is desirable to dampen reflected signals in order to reduce the background intensity below that required for easy conversation.

The intensity of reflected noise varies inversely with the room absorption. In this sense, the precise spatial effect or value of specific material changes can be estimated with the following formula:

$$NR = 10 \log A_2/A_1,$$

where NR is the noise reduction (in decibels); A_1 is the original room absorption (in sabins); and A_2 is the revised room absorption after material changes.

Ducts and Corridors

Sound absorbing materials are primarily intended to control noise within the room itself, not to control transmission between rooms. However, absorption is valuable in reducing transmission through air ducts and corridors. In these cases, extensive use of absorbing materials such as acoustic tile or sound-absorbing liners may be required to dampen transmitted noise.

Estimating Reverberation Time

The addition of sound-absorbing fabrics, upholstery, and other soft, porous materials will, of course, adversely affect the persistence of reflected sounds that are intended to reinforce the primary signal in lecture rooms, auditoriums, concert halls, and similar spaces that involve communication over significant distances.

As discussed in chapter 2, the problem of sound persistence and blending is a basic consideration in the development of appropriate signal reinforce-

Table 6-2. Sound absorption characteristics for selected typical interior surface materials.

	Selected Typical Absorption Coefficients (Hertz)					
	125	250	500	1,000	2,000	4,000
Typical wall materials						
High reflectance						
unglazed brick	0.03	0.03	0.03	0.04	0.05	0.07
plaster on brick	0.013	0.03	0.02	0.03	0.04	0.05
marble, glazed tile	0.01	0.01	0.01	0.01	0.02	0.02
concrete block, unpainted	0.36	0.44	0.31	0.29	0.39	0.25
concrete block, painted	0.10	0.05	0.06	0.07	0.09	0.08
smooth plaster on lath	0.013	0.015	0.02	0.03	0.04	0.05
rough plaster on lath	0.14	0.10	0.06	0.05	0.04	0.03
½-in. gypsum board panelling on studs	0.29	0.10	0.05	0.04	0.07	0.09
⅜-in. plywood paneling on studs	0.28	0.22	0.17	0.09	0.10	0.11
glass, typical window	0.35	0.25	0.18	0.12	0.07	0.04
glass, heavy plate	0.18	0.06	0.04	0.03	0.02	0.02
High absorption						
tackboard	0.42	0.49	0.33	0.22	0.19	0.17
lightweight drapery or tapestry (10-oz./sq. yd.), hanging flat on wall	0.03	0.04	0.11	0.17	0.24	0.35
mediumweight drapery (14 oz./sq. yd.), draped to half area on wall	0.07	0.31	0.49	0.75	0.70	0.60
heavyweight drapery (18 oz./sq. yd.), draped to half area on wall	0.14	0.35	0.55	0.72	0.70	0.65

Typical floor and indoor-outdoor ground materials

High reflectance

marble or glazed tile	0.01	0.01	0.01	0.01	0.02	0.02
concrete	0.01	0.01	0.015	0.02	0.02	0.02
tile on concrete	0.02	0.03	0.03	0.03	0.03	0.02
wood	0.15	0.11	0.10	0.07	0.06	0.07
water surface (pool or pond)	0.008	0.008	0.013	0.015	0.020	0.025

High absorption

heavy carpet on concrete	0.02	0.06	0.14	0.37	0.60	0.65
heavy carpet on padding	0.08	0.24	0.57	0.69	0.71	0.73
indoor-outdoor carpet	0.01	0.05	0.10	0.20	0.45	0.65
loose gravel-soil (4-in. depth)	0.25	0.60	0.65	0.70	0.75	0.80
grass (2-in. depth)	0.11	0.26	0.60	0.69	0.92	0.99
fresh snow (4-in. depth)	0.45	0.75	0.90	0.95	0.95	0.95

Typical floor-oriented furnishings and occupancy conditions

wooden pews, fully occupied (PSF)[1]	0.57	0.61	0.75	0.86	0.91	0.86
students seated in tablet-armchairs (PSF floor area covered)	0.30	0.41	0.49	0.84	0.87	0.84
leather seating, unoccupied	0.44	0.54	0.60	0.62	0.58	0.50
upholstered seating, unoccupied	0.49	0.66	0.80	0.88	0.82	0.70
upholstered seating, occupied	0.60	0.74	0.88	0.96	0.93	0.85

(continued)

Table 6-2 (continued).

	Selected Typical Absorption Coefficients (Hertz)					
	125	250	500	1,000	2,000	4,000
Typical ceiling materials						
High reflectance						
concrete	0.01	0.01	0.015	0.02	0.02	0.02
smooth plaster on lath	0.013	0.015	0.02	0.03	0.04	0.05
rough plaster on lath	0.14	0.10	0.06	0.05	0.04	0.03
5/8" gypsum board on studs or joists	0.29	0.10	0.05	0.04	0.07	0.09
3/8" plywood on studs or joists	0.28	0.22	0.17	0.09	0.10	0.11
recessed 2' × 4' fluorescent luminaire, plastic diffuser	0.30	0.22	0.17	0.10	0.10	0.10
(same, sabins/fixture)	2.4	1.8	1.4	0.8	0.8	0.8
High absorption						
(ASTM E-400 mounting)						
5/8" fissured tile (24 × 48 in.)	0.33	0.39	0.53	0.77	0.86	0.80
3/4" fissured tile (same)	0.57	0.60	0.65	0.83	0.94	0.98
1/2" glass fiber tile	0.75	0.80	0.85	1.0	1.0	1.0
typical absorption of air (sabins/1000 CF, 30% humidity)	—	—	—	—	3.5	11.5
Absorption of openings[2]						
large openings (typical)				0.50–1.00		
small openings (typical)				0.15–0.50		

1. PSF = per square foot of seating floor area.
2. Effect will vary depending on the absorption and volume on the other side of the opening.

192

ment. The objective is to produce full quality sound without excessive hardness or muddling. (See tables 2-4 and 2-5.)

In design, *reverberation time* is manipulated by altering room volume, the absorbing or reflecting characteristics of surfaces and finishes, and the density of room occupancy. This relationship is summarized in the following formula:

$$T_r = \frac{0.049 \ V}{\Sigma SA},$$

where T_r is the reverberation time (in seconds) — reverberation time is the interval required for a sound to decrease 60 dB, or to one-millionth of its original intensity, after the source has stopped generating; V is the room volume (in cubic feet); and ΣSA is the room absorption, including occupancy (in sabins).

In critical situations, reverberation time should be calculated for each significant frequency. However, for preliminary estimates, a single calculation for 500 cps may be sufficient.

Echo Control

Sound travels through room temperature air at a rate of approximately 1,140 ft/sec. This relatively slow speed means that there may be a significant time lag between the generation of a sound and reception at the ear of a distant listener. (If the distance is 100 feet, the time lag will be approximately 1/11 or 0.09 sec.)

When the ear receives only directly transmitted sound signals, the integrity of the signals is unaltered by any time lag. But if the direct sound is implemented by reflections from room surfaces, the various components may arrive at the ear at different times. If the time differential is short (approximately 0.035 sec or less for speech), the direct and reflected components reinforce each other. But a longer time lag may produce muddling interference that will reduce intelligibility.

Distinctly audible echoes will occur when a listener hears a sufficiently intense reflected signal 0.06 sec or more after he hears the direct signal. This condition is most prevalent in large spaces where the reflected sound path is at least 65 feet longer than the direct path (fig. 6-10).

In auditorium-type spaces, rear walls, ceiling areas adjacent to the front proscenium, or other surfaces that are remote from part of the audience may require special treatment to minimize or eliminate the reflected sound that produces the echo. This treatment will involve altering either the direction or intensity of the reflected sound. In this sense, commonly used techniques are: minimizing the area of the offending surface, and making the surface

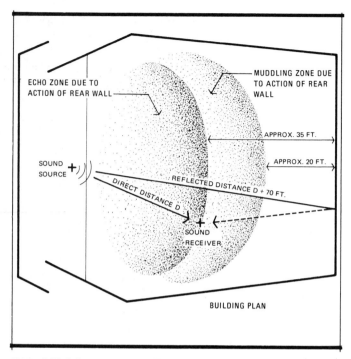

Figure 6-10. Influence of rear wall.

that remains highly absorbent; developing the offending wall as a diffusing surface, or tilting the surface to *ground* the reflection nearby; or a combination of both surface modulation and absorption.

Flutter Control

There is a similar effect in simple rectangular spaces (such as offices or gymnasium areas) where the walls are reflective and unbroken. In these spaces, multiple horizontal reflections may occur between a single pair of parallel walls. A sharp impact, such as hand clapping, will produce a ringing or buzzing sound that is called a *flutter echo*. This condition is most pronounced when the ear is at the same horizontal level as the sound source.

Flutter can be corrected by introducing draperies or similar absorbing materials on one of each pair of parallel walls. Alternative methods involve faceted wall panels or the use of diffusing objects such as large framed pictures, sculptural decoration, or venetian blinds.

CONTROL OF EXTERNALLY GENERATED NOISE

The signal-to-background ratio is also affected by noise that originates in adjacent or remote spaces. These may be airborne sounds that induce subtle vibrations in the separating wall, ceiling, or floor. Or, if the sound source is in direct contact with the physical barrier, the vibrations are induced through direct impact. Control involves the development of methods to reduce the intensity of induced vibrations, and the development of barriers that will effectively resist transmission of such energy.

Material Action

The installation of absorbing materials on the ceiling or walls of the receiving space will have only a marginal effect in limiting the penetration of external noise. In fact, the porous and lightweight character of most sound-absorbing materials seriously limits their value as sound insulators. Heavy materials are required, for the effectiveness of a sound-isolating barrier will depend on its weight or mass, its stiffness, and its air tightness.

Weight, Mass, and Stiffness

Heavy partitions provide higher inertia to resist sound vibrations. As a general rule, the greater the mass, the greater the reduction in sound transmission.

In general, the transmission loss characteristics of a partition will improve at higher frequencies. However, isolation is also a function of stiffness, for a very stiff partition will not resist transmission as well as might be expected on the basis of weight alone. In practice, large panels often exhibit a dip in transmission resistance over a limited range of frequencies. The precise location of this dip will depend on the size of the panel and its bending stiffness. For many partitions that utilize stiff, homogenous panels, however, this *resonance dip* can be as much as 10-15 dB in intensity and will occur within the speech range; thus permitting a higher than expected transmission of speech signals.

Air Space Separations

An isolating air space that separates individual partition layers will help to reduce the transmission of vibrations from one side of a barrier to the other. This method makes it possible to use lightweight construction methods while maintaining good resistance to sound transmission. It is also effective for reducing resonance dips. In both cases, care must be taken to insure that structural contact across the air space is minimized (structural discontinuity).

Transmission Loss

Wall and ceiling floor assemblies are rated according to their ability to resist the transmission of airborne sounds. The most commonly used method of rating is the *Sound Transmission Class* (STC), which utilizes the criteria curves discussed in chapter 2 (see fig. 2-10).

Figure 6-11 summarizes the performance of several typical assemblies. When interpreting these, note that normal speech can usually be heard clearly through a partition that is rated near STC 35. The same speech will be heard only as a murmur through an STC 40 wall, and will be negligible through an STC 50 wall. (Also see table 2-8.)

When more precise and specific analysis is required, transmission loss data is generally available for common construction details (table 6-3).

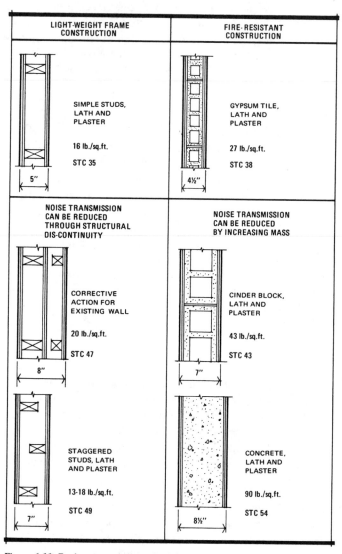

Figure 6-11. Basic construction principles.

Table 6-3. Sound transmission loss characteristics for selected interior barriers.

	Representative Construction Details	*Typical Transmission Loss (dB) @ Octave Band Center Frequencies, Hertz*					
		125	*250*	*500*	*1,000*	*2,000*	*4,000*
STC 25-30	¼″ single-pane glass	20	26	30	33	25	35
STC 30-35	Single wall: 2 × 4″ wood studs, ⅝″ gypsum board both sides (5.2 psf)	18	20	29	35	40	33
	double glazing, ¼″ glass, ½″ air space ¼″ glass	22	26	33	35	32	37
STC 35-40	single wall: 3⅝″ steel studs, ⅝″ gypsum board both sides (5.2 psf)	23	30	42	48	35	40
	6″ hollow core concrete block, painted both sides	33	35	36	41	48	54
STC 40-45	staggered 2 × 4″ wood studs, ⅝″ gypsum board, both sides (12 psf)	28	30	42	46	40	52
	3⅝″ steel studs ⅝″ gypsum board both sides, 2″ cavity insulation (6 psf)	26	40	51	57	53	45
STC 50-55	8″ poured concrete (or hollow core block filled with sand, 80-90 psf)	36	42	48	55	60	60
	3⅝″ steel studs, 2 layers ⅝″ gypsum board both sides, 4″ cavity insulation (11 psf)	34	47	56	60	58	52
STC 55-60	double wall: 2 wythes 6″ concrete block, 6″ air space, no bridging	38	47	58	78	85	59
	3⅝″ steel stud, resilient channel 1 side, 2 layers ⅝″ gypsum board both sides, 4″ cavity insulation	38	52	60	63	60	62

(continued)

Table 6-3 (continued).

Representative Construction Details	Typical Transmission Loss (dB) @ Octave Band Center Frequencies, Hertz						
	125	250	500	1,000	2,000	4,000
Typical floor-ceiling constructions							
STC 35-40 2 × 10″ wood joists, 16″ cc; ½″ plywood sub-floor, oak flooring, ⅝″ gypsum board nailed to joists[1]	17	23	28	36	47	42
STC 40-45 4 in. concrete slab; hardwood floor; plaster ceiling[2]	34	35	37	42	49	55
STC 50-55 2 × 10″ wood joists, 16 in. cc, ½″ plywood sub-floor, gypsum board ceiling on resilient channel, cavity insulation[3]	36	40	45	50	58	60
Typical doors							
1-¾″ hollow-core door, normally hung	10	15	16	18	21	22
1-¾″ solid-core door, normally hung	15	18	18	20	20	20
1-¾″ solid-core door, gasketed all edges	20	24	25	25	25	28

(Note: Transmission loss values will vary significantly with quality of crack closure at installation)

Data sources: Data from several acoustical laboratories and from manufacturers.

1. Normally acceptable for either airborne or impact noise; marginally acceptable for impact with heavy carpet and pad.
2. Normally marginally acceptable for airborne noise; marginally acceptable for impact noise with carpet and pad.
3. Normally acceptable for airborne noise; acceptable for impact noise with carpet and pad.

Estimating the Effect of Composite Barriers

Most walls include more than one type of construction. This difference may involve a door or window section that has a noise reduction value (TL_o) below the value assigned to the greater portion of the barrier (TL_w). (If the minor area is a hole or opening, $TL_o = 0$.) The effective or average transmission loss of the composite wall (TL_e) can be estimated from figure 6-12.

Estimating Noise Control Value (External Noise)

The effective transmission loss of the barrier is the most significant single factor in determining the intensity of transmitted sounds. To a lesser extent, however, intensity is also affected by the area of partition surface and by the room absorption of the receiving room. In this sense, the precise spatial effect in the receiving room can be estimated with the following formula:

$$NR = TL_e - 10 \log S/A_2,$$

where NR is the noise reduction (in decibels); TL_e is the effective transmission loss of the homogeneous or composite wall (in decibels); S is the area of

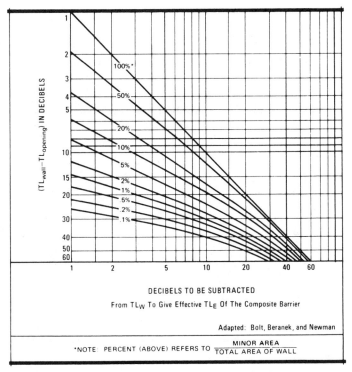

Figure 6-12. Effective transmission loss of a composite barrier.

common surface between the *source room* and the *receiving room* (in square feet); and A_2 is the room absorption of the *receiving room* (in sabins).

Note that if the room absorption (A_2) is low relative to the surface area (S), the effect of the receiving room is to diminish the transmission loss value (TL_e). However, if the room absorption numerically exceeds the surface area, the effect of the receiving room is to increase the total noise reduction.

Flanking Noise

No matter how good an insulator a partition may be, cracks and holes can very quickly nullify its value. Obviously the design and placement of doors and windows are major considerations. But poorly designed ceiling cavities, wall-to-ceiling or wall-to-floor cracks, back-to-back switch boxes, and similar paths can adversely affect the insulation value of a partition (fig. 6-13). These paths must be designed to minimize sound transmission and leakage. To avoid flanking sound from one private office to another, the wall/ceiling juncture must be well-treated. Figure 6-14 illustrates three techniques that can be used to eliminate flanking sound problems.

Structural Impact

When an intermittent sound source comes into direct contact with the barrier (such as occurs with footsteps or simple hammer action), transmission is reduced by inserting a resilient material between the source and the barrier.

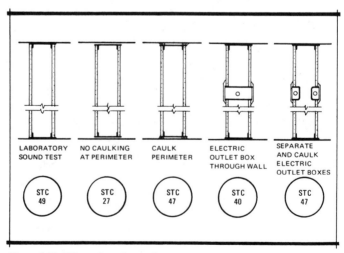

Figure 6-13. Effect of crack openings.

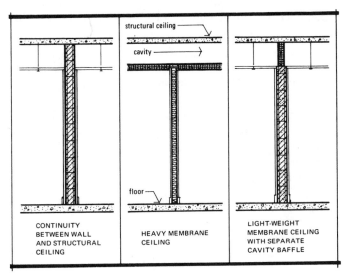

Figure 6-14. Methods of cavity closure.

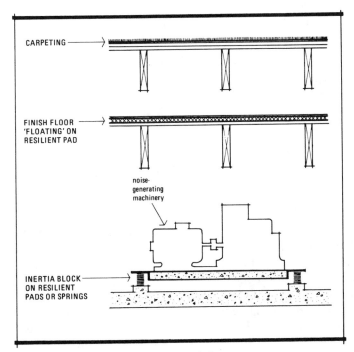

Figure 6-15. Reduction of impact transmission.

Table 6-4. Impact sound insulation for selected typical floor–ceiling systems.

System Type	Typical IIC
Reinforced 4-in. thick concrete slab with tile floor above	IIC-29
Reinforced concrete slab with wood floor over resilient layer	IIC-45
Reinforced concrete slab with cement topping floor and suspended gypsum plaster ceiling	IIC-47
6-in. precast concrete slab with cement topping floor and plaster coat ceiling (7½-in. total thickness)	IIC-30
5-in. precast concrete joists with wood floor on wood battens over resilient pad and cement coat and gypsum board ceiling on wood battens	IIC-53
2 × 8 wood joists, 16-in. o.c., wood floor and gypsum board ceiling	IIC-32
Add carpet and pad to 2 × 8 wood joist system	IIC-56
Add insulation between wood joists and resilient mounting of gypsum board ceiling	IIC-70
14-in. steel joists, 2½-in. thick concrete topping, carpet and pad with plaster ceiling	IIC-64
14-in. steel joists, 2½-in. thick concrete topping, tile floor with plaster ceiling	IIC-35

SOURCE: HUD: *Airborne, Impact, and Structure Borne Noise*

Structural transmission may also involve sustained vibrations (such as fan vibrations or vibrations caused by a rotating or reciprocating engine). When such vibrations are subtle, the insertion of resilient padding may again be sufficient for control. For more intense vibration, springs may be inserted between the source and the barrier (fig. 6-15).

As a simplified procedure for preliminary estimating purposes in general-use spaces, floor assemblies are rated according to their ability to resist the transmission of more typical impact noises. This is the Impact Insulation Class (IIC) which is discussed in chapter 2. Table 6-4 summarizes the performance of several typical assemblies.

MECHANICAL NOISE

Noises that are associated with mechanical equipment can be classified in several categories.

1. *Airborne equipment noise that is transmitted to occupied spaces through common walls or ceiling-floor assemblies:* This is primarily controlled through partition or barrier design, to insure sufficient mass, minimum stiffness, and possibly structural discontinuity (air space separations). All joints and cracks must be effectively sealed.

2. *Airborne equipment noise that travels through air supply and return ducts or plenums:* Generally, this is controlled by placing absorbing blankets and linings within the duct or plenum. However, particularly for short duct lengths and wall openings, special silencers or mufflers may be required (fig. 6-16).

3. *Structural-borne sounds induced by vibrating equipment that is rigidly attached to the structure or barrier:* This is generally controlled through the use of rubber mounting pads or steel springs. Note that it may be necessary for these devices to absorb deflections of up to 6 inches for powerful low-RPM equipment.

A related method of treatment involves hanging the equipment from the ceiling with vibration hangers.

Care must be taken to isolate all paths to and from the equipment, including the provision of resilient or flexible couplings for pipes, ducts, and conduit.

4. *Internal noise induced by air supply diffusers and grilles that are located within the room itself:* The rated noise level of the mechanical device itself is usually the responsibility of the manufacturer. So this aspect is primarily controlled by preparing a suitably rigid performance specification

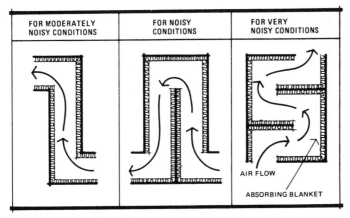

Figure 6-16. Sound "traps."

Table 6-5. Estimation of acoustical performance (intensity).

	Formulas	References
Noise reduction room absorption	$\Sigma SA = S_1A_1 + S_2A_2 + S_3A_3 + \ldots$	See tables 2-1 and 6-2 for criteria
wall and ceiling barriers (transmission)	Obtain curve of transmission loss (or STC rating)	See tables 2-8 and 6-3 and fig. 2-8 for criteria
floor-ceiling barriers (impact)	Obtain curve of impact insulation class (or IIC rating)	See tables 2-9 and 6-4 and fig. 2-9 for criteria
internal operating equipment	Obtain curve of operating equipment noise (or NC rating)	See fig. 2-7 and tables 2-6 and 2-7 for criteria
Signal reinforcement natural amplification	Develop ray diagrams	See fig. 6-6 and list under *Natural Sound Reinforcement*
control of echoes and muddling	Analyze ray diagrams	See fig. 6-10
reverberation time	$T_r = 0.049V/\Sigma SA$	See tables 2-4 and 2-5 and fig. 2-5 for criteria

to guide equipment selection (see chapter 2 for a discussion of *Noise Criteria Curves*).

When suitably rated devices are not available, air noise can be further controlled by reducing large pressure drops within the distribution system. Noise reductions can be realized, then, by selecting an oversized unit with the intention of operating it below normal rated capacity in order to reduce the sound level.

A second alternative is to select several smaller units. When this is done, however, care should be taken to insure that the cumulative noise effect does not exceed the applicable Noise Criteria Curve. (See fig. 6-7 for estimating the additive effect of multiple noise sources.)

Calculated and Graphic Estimates of Acoustical Performance

When estimating the comprehensive acoustical performance of a space, table 6-5 is useful for summary purposes.

CHAPTER 7

Thermal and Atmospheric Control

HEAT
Definition

Heat energy is generated by the random motion of molecules, and the intensity of this molecular action is measured by thermometer scales. A temperature of *absolute zero* ($-459°$F or $-273°$C) is the point at which heat is no longer being generated because all molecular motion stops. The unit measure ($1°$F; $1°$C) relates to the quantity of heat energy, defined as a Btu or calorie, required to raise a given amount of water one step on the scale (one Btu raises one pound of water from $60°$F to $61°$F; one calorie raises one gram of water from $15°$C to $16°$C).

As the intensity of molecular action increases, the molecular structure becomes more diffused and the matter will assume different physical forms. Generally, this is a progressive change from a *solid* to a *liquid* (water changes at $32°$F) to a *gas* (water changes at $212°$F).

Heat is analogous with liquids and electricity in the sense that a difference in intensity (or pressure) produces the potential for flow. Heat always flows from a warm object or mass to a cooler one (that is from the higher rate of molecular action to the lower rate). This flow of energy continues until a neutral or equal temperature condition exists between the two. An energized electric lamp, for example, is warmer than the surrounding air and warmer than the body surface of a human being. Since thermal gradients exist, a flow of heat results. Under extreme conditions, this heat can produce discomfort for the occupant.

Heat Transfer

Heat transfer related to human comfort takes place in three different ways:
• *Radiant heat* is short-wave infrared energy that passes as a ray or beam through air or any transparent medium with little absorption or dispersion.

When energy in this electromagnetic form strikes an opaque object such as a person or an environmental surface, heat is absorbed and reflected in a manner similar to the action of light, and a rise in surface temperature results due to the absorbed energy. This form of heat is difficult to control with conventional air circulation methods because the transfer of heat is directly between the source and the receiving surface or object, and the surrounding air exerts little influence in this process.

Control of heat in this form, then, depends on techniques that manipulate the temperature gradients between surfaces (or between surfaces and energy sources). This involves techniques that adjust the overall relationship of surface temperatures within an enclosed environment, relating the temperature of these surfaces to the surface temperature of the human skin. Also, techniques that modify the directional concentration of radiant energy such as solar reflectors and solar shading devices can be used to control radiant energy.

• *Conduction/convection* is the second type of heat transfer and it does respond to conventional air circulation methods. This is the most prominently used method of environmental control for either heating or cooling in today's built environments. Heat is transferred directly from a hot object to the surrounding air by *conduction;* the warmed air then moves by gravity *convection* (hot air rises) or by force convection to a cooler object, where heat is again transferred by conduction. This last transfer warms objects and surfaces within the environment.

Heat flow is reversed in an environmental cooling system with heat going from the warmed air to a cooled medium.

The general characteristics of convective motion are involved whenever heat is borne by a fluid such as liquid hot water, warmed or cooled environmental air, or a gas, such as steam. The rate of conductive heating or cooling depends on the velocity of the circulating medium and on the difference in temperature between the medium and the surface to be heated or cooled.

• *Evaporation* is an increasingly important controlling factor when environmental temperatures are high. Although this method of heat transfer is simply the change of state of water produced by the primary heat transfer methods of radiation and conduction/convection, it is a critical environmental heat transfer concept. In fact, at high temperatures, air and most untreated room surfaces tend to lose their cooling potential, and the human body will depend almost completely on evaporation in its effort to maintain thermal equilibrium.

Evaporation of surface moisture (such as perspiration) will cool any object or body because heat is taken from the mass to facilitate the evaporative process.

For this reason, control of water vapor in the environmental air is an important qualitative aspect. Humidity should be low enough in warm weather to facilitate evaporative cooling of the skin, and high enough in dry weather to prevent excessive drying or dehydration.

These three forms of heat transfer affect and are the means for manipulation of the thermal environment for human comfort and experience. The habitable thermal environment is partially achieved through the development of an enclosing building shell; and it is further balanced with heat-generating and heat-consuming devices. Both concepts must be evaluated in the context of the natural site and climatic environment.

THE SOLAR ENVIRONMENT

The earth revolves about the sun in an approximately circular path, with the sun located slightly off-center in this circle. The earth is nearest to the sun about January first and is at the most remote position about July first.

As a result of this slightly eccentric position of the sun, the intensity of solar radiation at the outer limit of the earth's atmosphere is about 7 percent higher on January first (approximately 445 Btu/sq ft/hr) than it is on July first (approximately 415 Btu/sq ft/hr). This represents a fluctuation of 3.5 percent from the *mean* solar radiation at the outer limit (generally taken to be 429.5 Btu/sq ft/hr).

As this radiation passes through the atmosphere, part of the energy is reflected and scattered by dust particles and by water and air molecules. This scattered energy becomes the diffuse sky vault. In addition to this scattering effect, part of the solar radiation is absorbed by ozone in the upper atmosphere and by water vapor near the surface.

The depletion of this ozone layer in the outer atmosphere allows excess ultraviolet radiation to reach the surface of the earth. This possible depletion is the subject of much investigation and is generally connected with the production and release of chlorofluorocarbons (CFCs). Molecules of this inert substance eventually reach the outer atmosphere where they break down the ozone molecules.

Atmospheric Variables

The energy that is incident at the surface of the earth is therefore present in two forms: (1) direct radiation that has been somewhat diminished in its passage through the atmosphere and (2) diffuse sky radiation.

In this respect, as the atmosphere becomes congested with clouds, dust, or smoke and other industrial contaminants, the direct radiation is increasingly diminished. For example, on a clear day at sea level, the proportion of direct radiation to diffuse radiation is typically 70:30. In many large cities, where the atmosphere is contaminated by industrial gases and particles, this ratio will change. In this situation, a clear day ratio of 45:55 or less is more typical.

On overcast days, the diffuse component will increase further, with a corresponding decline in direct radiation; until, with heavy cloudiness, nearly all of the incident energy is in the diffuse, nondirectional form.

In addition to the diffusing effects of clouds and contaminants, intensity changes are also inherent with clear sky conditions. This is due to the effect provided by the depth of the air mass through which the rays must pass. This depth of passage is thinnest at noon in summer, where the sun is nearly overhead. In this situation, an approximate 294 Btu/sq ft/hr is incident on a surface normal to the plane of the sun's rays. At 3 P.M. on the same day, the slightly more oblique passage of energy (caused by rotation of the earth relative to the sun) reduces the maximum at the surface to 266 Btu/sq ft/hr. At 6 P.M., passage through the atmosphere is much more oblique and the maximum amount which penetrates to the earth's surface is reduced to 67 Btu/sq ft/hr.

Geographic and Climatic Variations

If the earth is evaluated as a thermodynamic system, with outer space as the external environment, radiation is the only form of heat transfer that can maintain an overall heat balance. It is estimated that 43 percent of the solar energy that is incident at the outer limit of the atmosphere reaches the earth's surface. After it absorbs radiant energy, the earth itself becomes a secondary radiant source, emitting energy toward outer space.

The blockage of this earth-emitted radiation by the atmosphere is what is typically described to be the *greenhouse effect*. The burning of fossil fuels has contributed to the build-up of atmospheric contaminants which block earth-emitted radiation. Also, the depletion of the earth's forests has decreased the rate of removal of carbon dioxides from the earth's atmosphere by photosynthesis.

Earth surface solar gains and losses will differ considerably with local conditions. The most significant of these conditions relates to the variable and nonuniform distribution of solar energy that is caused by the 23.5 degree tilting of the axis of the earth. This produces *seasonal variations* in the altitude of the sun above any specific location on earth. During winter periods, it also exaggerates the fact that the more northern and southern latitudes receive sun rays at an angle of incidence that is much more oblique than it is at the equator. As noted previously, this reduces significantly the amount of penetrating solar energy.

Because of these effects, the portion of the earth's surface that is defined to be between 30 degrees North latitude and 30 degrees South latitude exhibits a net heat gain by radiation, while the two segments between 30 degrees latitudes and the poles exhibits a net heat loss by radiation.

Without atmospheric circulation, therefore, the regions near the equator would become progressively hotter, while the areas of higher latitude (north and south) would become progressively cooler. The general wind patterns and ocean currents provide the primary means for transporting heat from the equator toward the poles.

Atmospheric Heat Storage

When the earth emits energy toward outer space, some of this energy penetrates the atmosphere and escapes. But the greater portion (approximately 70 percent) is intercepted by water vapor in the atmosphere and is again returned to the earth by radiation or convection. On overcast days, the clouds form a local barrier that minimizes losses from the earth's surface. Thus, the water vapor in the atmosphere exerts a greenhouse effect, retaining heat and stabilizing temperatures at the surface of the earth.

If the barriers are changed significantly, the thermal balance of the earth could be severely affected. A build-up of wastes in the atmosphere could increase the temperature of the earth.

Together with the thermal storage capacities of the earth, including both land and water masses, the atmospheric barriers provide a stabilizing effect that is somewhat seasonal. The peak atmospheric air temperatures lag behind the solar conditions. Heat storage effects that produce maximum summer temperatures in most of the United States do not generally prevail until several weeks after the maximum solar gains which occur about June 21, when the sun reaches its most northerly position. Sustained peak temperatures generally occur in early August.

Sustained minimum temperature conditions similarly occur in late January or early February—several weeks after the minimum solar energy condition which occurs about December 22, when the sun reaches its most southerly position.

For design purposes, four approximate dates are necessary to determine the conditions of maximum heat gain and maximum heat loss for a given location. They are: the dates of maximum solar radiation, maximum air temperature, minimum solar radiation, and minimum air temperature.

Earth Surface Variables

If the earth were a flat surface with no change in elevation or without bodies of water, the climate of a given location would directly correspond with its latitude. But any site can experience diverse microclimatic conditions. Factors to consider are:

• *Elevation:* The higher the elevation of a site, the cooler the ambient temperatures will be. Air temperature drops at a rate of 2.5° to 3.0° per 1000' increase in elevation.

• *Exposure to sun and prevailing winds:* South-facing slopes in the northern hemisphere will gain significant amounts of solar radiation while north-facing slopes will gain minimal amounts of solar radiation. East and west slopes will gain solar radiation in daily and seasonal cyclical patterns, less than south slopes but more than north slopes. Similar analysis is necessary for the impact of wind on a site. Local and regional patterns must be ascertained to determine the prevailing and seasonal wind impacts upon a site.

• *Topography:* Cold air will tend to drop to low areas and create air movement through valleys and around hills and mountains. Solar radiation will strike slopes differently depending upon orientation and time. Consequently, warmed air currents and wind patterns will be created.

• *Size, shape and proximity of water bodies:* Water requires approximately four times as much heat to raise its temperature as does an equivalent unit mass of earth. Conversely, it must lose four times as much heat per °F to lower its temperature. Consequently, the temperature of water changes much more slowly than that of land. The net effect of the presence of water is to moderate diurnal and seasonal temperature extremes. The larger the body of water, the more pronounced is this effect.

• *Soil structure:* Dry soils will tend to hold heat for longer periods of time than moist soils. The fluids in moist soils will warm and cool in response to the adjacent surfaces, and thus begin to circulate and transport heat. The result of the heat transport is to neutralize temperature differentials. For this reason, underground buildings which do not experience large temperature differentials still need insulation so as to prevent heat transfer to surrounding soils.

• *Vegetation.* Trees, shrubs, meadows, and crops will reflect and absorb solar radiation differently. An analysis of the ground cover and the corresponding solar reflectances will be necessary for each site.

• *Man-made structures.* Buildings, streets, parking lots, and lawns will create artificial microenvironments similar to the effects generated by the above-mentioned topography and vegetation.

THE BUILDING MICRO-CLIMATE

Buildings are, in part, a response to the variable solar and climatic influences on earth. The aspect of thermal comfort is a significant criteria in the development of the building enclosure. The term *comfort* can be broadly defined as the condition where thermal stress is minimized and where the occupant can adjust his or her environment with a minimum of automatic body effort. The thermal function of the enclosure is to assist in the establishment of an area of biological equilibrium.

Regional Building Form

In its basic form, the building enclosure functions to promote heat gain and conserve heat when the external environment is cold, and to impede

Figure 7-1a. Example of regional building form for polar climate: the igloo (photo: National Museums of Canada, negative J4439).

heat gain and dissipate heat when the external environment conditions are warm. There are four basic climatic zones with corresponding typical built forms that express a response to the regional thermal environment. These are polar, hot arid, tropical, and temperate.

Polar

The thermal built-form function is most readily observed in the more extreme climatic regions. For example, the domed *igloo* (fig. 7-1a) by the Eskimos is a response to the prolonged and extreme cold weather conditions in areas where materials are scarce. This snow-packed hemispherical form is compact (a minimum enclosure surface); it is well sealed and insulated by the snow; and the form effectively deflects winds, with entrance openings oriented away from the prevailing winds to reduce drafts.

Many subartic structures are somewhat similar in concept in that they have common party walls, thus reducing the exposed envelope surface area, and minimum-surface exterior shapes. These characteristics combine to

Figure 7-1b. Example of regional building form for hot arid climate: adobe dwellings (photo credit: Tyler Dingee; courtesy of Museum of New Mexico, negative 91966; Palace of the Governors, Santa Fe, New Mexico).

reduce the exposed surface area of each habitable unit. There are also variations of double-shell construction techniques which provide an effective insulating barrier with minimum surface area.

Hot, Arid

At the other extreme, it is interesting to note that hot and arid regions (semidesert) have tended to develop somewhat similar enclosure concepts. The Pueblo Indians developed large communal structures that utilize party walls to reduce exposed surface area (fig. 7-1b). They were not overly concerned about insulation, but massive adobe walls acted to delay heat transmission while small window and door openings were placed to minimize solar exposure.

Domed roof structures are also common in these hot, arid regions. This prevalence is a response to the intense and unshaded solar loads that are incident on roof areas in semidesert regions. When a dome is substituted for

Figure 7-1c. Example of regional building form for tropical climate: Dogtrot House (photo: Professor Arnold Aho).

a flat roof, the roof area is several times the base area. This spreads the incident solar energy over a large area, reducing its average density. The thermal action is further implemented by winds that tend to dissipate heat from the enlarged surface. Where masonry materials store heat by day, night breezes will help to dissipate that heat before it penetrates to the interior.

Tropical

Each of the extreme climate regional building forms evolved from the need to control (1) convection-conduction transfers through the enclosure and (2) radiation transfers to and from the occupants caused by cold or warm peripheral enclosure surfaces. A building design in hot and humid regions (such as the tropics) is faced with a very different problem. In these regions, evaporation of moisture for dampness control and for occupant cooling effect is very important. In tropical areas, habitable structures have minimum walls to make maximum use of penetrating breezes for evaporation control (fig. 7-1c). Furthermore, floors may be elevated to facilitate ventilation. Tree cover and roof structures are then developed to provide shade and insulation. Roof overhangs are developed to protect the minimum walls against rain and solar radiation.

Figure 7-1d. Example of regional building form for temperate climate: traditional style with climatic variations (photo: Professor Foster Armstrong).

Temperate

It is in the temperate regions, which include much of the United States, that the greatest potential for structural diversity exists. Since these zones produce extremely variable climatic conditions, the building may periodically be called upon to respond in a manner similar to each of the preceding regional categories; at times providing an effective barrier to exclude extreme cold or extreme heat, at other times providing shade plus openness to permit the penetration of cooling breezes. This conflict of criteria to meet temporary and seasonal changes in the basic external environment has produced the considerable diversity of form that is found in these temperate regions. It has also been the stimulus to develop mechanical devices to compensate for the compromises that must inevitably be made to building form, materials, and orientation of openings.

This conflict of criteria requires that builders in temperate regions distinguish between periods when one's goal is to seek solar heat gains, such as in cold weather, and periods when one seeks to avoid solar gains, such as in hot weather. Similarly, there are periods when winds are beneficial and periods when winds should be intercepted. Building orientation, materials, and forms are often a compromise with mechanical devices utilized to compensate for the conflicts.

The traditional suburban residential unit expresses this compromise as a minimal exterior envelope modified by opportunities for seasonal openness through elements such as the porch, fenestration, and roof openings such as dormers and skylights (fig. 7-1d).

Heat Transfer through the Building Envelope

Several factors influence an attempt to develop satisfactory enclosures that respond to the varying solar and climatic conditions. Analysis of these factors falls into the following general categories:

- Instantaneous heat gains associated with solar energy that passes through transparent openings in an enclosure shell.
- Delayed heat gains associated with solar energy that impinges on opaque portions of the shell.
- Heat gains or losses associated with conductive transfer through the shell to or from adjacent outdoor air.
- Heat gains or losses associated with infiltration of air through cracks and openings in the shell.
- Heat gains or losses associated with intentional expulsion of stagnant or contaminated indoor air and replacement with fresh outdoor air.

Penetration of Solar Radiation through Openings

The transmission of solar radiation through openings in the enclosing shell is nearly instantaneous. Yet the cooling load created by solar radiation is not instantaneous. In fact, it is common for cooling requirements in buildings to peak several hours after the sun has passed through its prime position for introducing solar radiant energy into a structure. This is due to the fact that the opaque mass of a building which receives the energy will absorb and store it for a period of time. For winter heating conditions this effect can be advantageous as the heat of the sun is initially stored and then reradiated into the interior as the environment cools.

The intensity of the solar transmission at a given point and time will depend on the orientation of the openings and the reflection, absorption, and transmission characteristics of the transmitting material (fig. 7-2).

For roof openings, such as a skylight, the atmospheric air mass is thinnest (and therefore offers the least resistance) when the sun is directly overhead. This is the time when rays strike the horizontal roof at something approaching normal incidence; so direct solar radiation through these openings approaches a daily maximum at solar noon on June 21.

Vertical wall openings rarely receive direct radiation at near-normal incidence. The more oblique passage of rays through the atmospheric air mass tends to reduce intensity. This means that radiation transmitted through windows

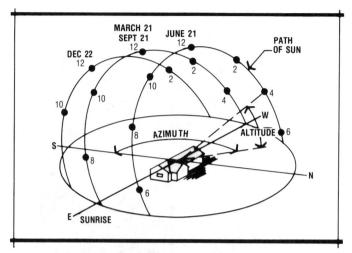

Figure 7-2. Azimuth and altitude angles: the location of the sun relative to the position of the building being analyzed. Azimuth angle: horizontal ground plane angle between due south (0°) and the plan position of the sun. Measured in a counterclockwise direction during A.M. hours and in a clockwise direction during P.M. hours. Altitude angle: vertical angle between the horizontal ground plane and the position of the sun.

and doors is generally lower in maximum intensity than that transmitted through roof skylights. See figure 7-3 for an example of solar position at 40° North latitude and tables 7-1 and 7-2 for representative solar heat gain values.

The more significant thermal control problems created by enclosure openings occur when intensity of solar radiation is near maximum. At these times, direct rays must generally be tempered in the immediate vicinity of the human occupant. This can be accomplished in two ways: (1) by orientation of openings to minimize the penetration of direct solar radiation or (2) by shading those openings that must remain exposed to direct radiation at a time when this may become an adverse influence on comfort or basic system performance.

In addition to the thermal control problems, the daylighting glare problems must be considered when designing transparent openings. Direct sunlight is too bright for human visual tasks within the context of the interior environment. This can be controlled by reflecting the direct sunlight from appropriately located surfaces and thus diffusing the light so as to decrease its intensity. Aalvar Aalto's development of daylighting techniques exhibited these principles by keeping the source of direct sunlight out of the visual field of the occupants of the space. Daylight comes in high and is seen primarily as reflected light from the interior spaces.

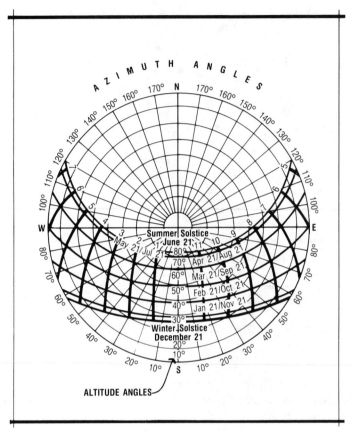

Figure 7-3. Solar path diagram, 40° North latitude. Representative graphic view of the path of the sun as projected on the horizontal ground plane. Altitude and azimuth angles can be read for each month and hour of the year. Altitude angles can be read on the equally spaced 10° interval concentric circles. Azimuth angles can be read on the equally spaced 10° interval radii.

Orientation of Walls and Openings to Respond to Solar Load

For many combinations of orientation and latitude, a wall opening will receive only grazing radiation. More critical are surfaces which receive radiation at more nearly perpendicular angles. As an example, in the northern temperate zone:

- Walls facing south will receive maximum direct radiation in the winter, early spring, and late fall.
- Walls facing east will receive maximum radiation at sunrise and during the early morning hours.

Table 7-1. 40° North latitude: Approximate solar heat gain incident on the outside of building surfaces (Btu/sq ft/hr).

40°	March 21			June 21		
	8 AM	*Noon*	*4 PM*	*8 AM*	*Noon*	*4 PM*
Overcast day	6	16	6	21	36	21
Clear day						
direct normal	250	307	250	246	279	246
N	16	29	16	30	38	30
NE	91	29	16	156	38	27
E	218	31	16	216	41	27
SE	211	145	16	152	72	27
S	74	206	74	29	95	29
SW	16	145	211	27	72	152
W	16	31	218	27	41	216
NW	16	29	91	27	38	156
horizontal roof	85	223	85	153	267	153

40°	September 21			December 21		
	8 AM	*Noon*	*4 PM*	*8 AM*	*Noon*	*4 PM*
Overcast day	8	20	8	1	6	1
Clear day						
direct normal	230	290	230	89	285	89
N	17	30	17	3	18	3
NE	87	30	17	8	18	3
E	205	32	17	67	19	3
SE	199	142	17	84	178	3
S	71	200	71	50	253	50
SW	17	142	199	3	178	84
W	17	32	205	3	19	67
NW	17	30	87	3	18	8
roof	82	215	82	6	113	6

New York City (40°−47′); Pittsburgh (40°−27′); Columbus (40°−0′); Chicago (41°−53′); Kansas City (39°−7′); Boulder (40°−0′); Salt Lake City (40°−46′); Redding, CA (40°−31′).

• Walls facing north will receive radiation only in the late spring, early fall, and summer in the early morning and late evening.
• Walls facing west will receive maximum radiation during the afternoon hours or at sunset, particularly during the summer months.

Openings in the east and west walls are therefore subject to direct radiation loads that vary considerably during a typical day. These openings permit particularly severe penetrations early and late in the day because low sun

Table 7-2. 32° North latitude: Approximate solar heat gain incident on the outside of building surfaces (Btu/sq ft/hr).

32°	March 21			June 21		
	8 AM	*Noon*	*4 PM*	*8 AM*	*Noon*	*4 PM*
Overcast day	7	18	7	20	37	20
Clear day						
direct normal	260	313	260	245	280	245
N	17	32	17	36	41	36
NE	107	32	17	171	41	27
E	227	33	17	214	42	27
SE	209	122	17	135	52	27
S	62	176	62	28	60	28
SW	17	122	209	27	52	135
W	17	33	227	27	42	214
NW	17	32	107	27	41	171
roof	100	252	100	151	276	151

32°	September 21			December 21		
	8 AM	*Noon*	*4 PM*	*8 AM*	*Noon*	*4 PM*
Overcast day	9	22	9	2	9	2
Clear day						
direct normal	240	296	240	176	304	176
N	18	33	18	7	22	7
NE	103	33	18	19	22	7
E	215	35	18	135	23	7
SE	198	120	18	166	177	7
S	60	171	60	97	252	97
SW	18	120	198	7	177	166
W	18	35	215	7	23	135
NW	18	33	103	7	22	19
roof	96	244	96	22	158	22

Jacksonville, FL (30°−30′); Charleston, SC (32−54′); Atlanta (33°−39′); New Orleans (30°−0′); Dallas (32°−51′); Tucson (32°−7′); San Diego (32°−44′).

angles create conditions in which the sun's rays are nearly perpendicular to the plane of the openings. Furthermore, with west walls, peak solar gains occur coincidentally with high afternoon air temperature and humidity conditions during summer periods. This means that west openings admit heavy solar gains at the same time that peak ventilation, infiltration, and conduction heat gains occur.

The nature of these maximum radiation loads suggests that buildings in the northern temperate regions are generally best developed so that most major openings are toward the south. Control of summer solar radiation is most easily achieved through horizontal roof overhangs and louvers. The same opening can admit the lower altitude winter sun for passive heat gain during cold weather. East openings provide early morning radiation with the benefits of thermal tempering and early sunrise stimulus. North openings provide uniform diffuse daylighting with minimum direct radiation. The west exposure is the least desirable due to summer thermal and glare problems.

Shading Techniques and Devices

Direct solar radiation constitutes one of the most significant thermal influences on the building shell. Because the character of this load changes with time of day and season of the year, perimeter locations within the enclosure will constantly undergo thermal changes, and occupant comfort may be adversely affected on a localized basis.

The principal thermal problem here derives from the fact that clear glass permits almost unimpeded penetration of radiant energy. To minimize extreme variations in thermal performance among openings and walls of differing orientation, therefore, some assemblies may be modified to intercept direct radiation and minimize penetration into the occupied interior.

Measurement of the reduction in transmitted radiation is in terms of the shading coefficient. The *shading coefficient* is the ratio of:

$$\frac{\text{Total solar heat gain with shading (Btu)}}{\text{Total solar heat gain without shading (Btu)}}$$

The shading coefficient of unshaded and untreated "double strength" ⅛-inch plate glass is 1.0, making this the base condition of comparison. It includes consideration of direct transmission, absorption, and reradiation (table 7-3).

Treatment can take several forms. The basic alternatives that follow are listed in increasing order of reliability, effectiveness, and preference.

Interior Blinds and Shades

While effective for shading room occupants from direct radiation, interior shading devices permit radiant heat to enter the space itself before it is intercepted and absorbed. As a result, the absorbed energy may become an adverse thermal influence through interior convection and reradiation. Light-colored finishes will, however, increase the effectiveness of the shading system because a greater proportion of the incident energy is reflected back through the glass to the exterior.

Table 7-3. Shading and "U" coefficients for windows.

| Glass Type | Inside Shade | | | | | |
| | None | | Drapery or Venetian Blind | | Opaque Shade | |
	SC	U	SC	U	SC	U
Single	1.00	1.11	0.50	0.81	0.38	0.81
Double	0.88	0.57	0.45	0.55	0.36	0.55
Heat absorbing	0.58	0.45	0.37	0.44	0.33	0.44
Triple	0.80	0.38	0.44	0.40	0.36	0.40

SC is percent of light transmitted relative to single glass. U is in Btu/hr-sq ft-°F.
SOURCE: ASHRAE, 1989 *Handbook of Fundamentals.*

Absorbing and Reflecting Glass

Clear glass can be replaced with reflecting or absorbing glass. The absorbing glass holds a portion of the radiation for dissipation by convection and reradiation both to the exterior and the interior. It is not as effective as reflecting or "low-E" glass which reflects the radiation back to the exterior before it penetrates to the interior. The color of the reflecting coat plays a significant role in the percentage of radiation reflected. Different coatings possess varying capabilities in reflecting the total solar spectrum as opposed to the ability to reflect just the visible spectrum.

Landscaping and Adjacent Structures

Natural and man-made elements appropriately placed in the exterior environment can be effective barriers when preserved in a useful condition. Deciduous trees are particularly effective, as they correlate with the seasonal cycles of the earth.

External Architectural Treatment

External shading devices can be architecturally expressive of the regional nature of the building as they respond to the local conditions. They are more effective than similar interior devices because energy interception occurs outside of the occupied space where it can be dissipated with minimum effect on the interior environment.

Architectural shading can take several forms. Shading can be provided by functional elements such as balconies or plan setbacks; it can be provided by structural extensions such as overhanging cantilevers; or it can be provided by secondary structures such as awnings, screens, or fixed louvers. Figure 7-4 illustrates a variety of external architectural devices. In cold climates, these

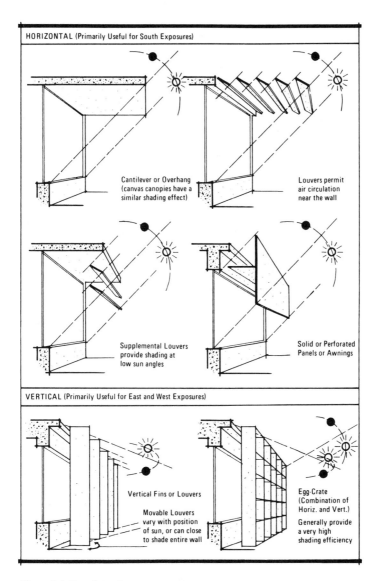

Figure 7-4. Exterior solar screens.

devices should allow for winter solar penetration while shading the solar energy in the summer.

Penetration of Solar Radiation through Opaque Assemblies

Opaque assemblies provide traditional types of enclosure systems which can facilitate the provision of thermal insulation and the rejection of solar radiation. It does so with the loss of the opportunity to provide view, daylighting, and the capture of beneficial solar radiation.

When solar radiation is incident on an opaque enclosure or shell, part of the incident energy is reflected (table 7-4), a relatively small part is transmitted, and the remainder is absorbed. Since this absorbed energy remains within the assembly until it is reemitted from either the exterior or the interior surface, the heat storage properties of the enclosure become a potentially significant consideration in transmission analysis.

This heat storage capacity will tend to produce a time lag—that is, the interior thermal environment will tend to lag behind the external weather conditions. The magnitude of this time lag will depend on the storage capacity of the enclosure, with the lag generally lengthening as material density or mass increases (table 7-5).

If the enclosure storage capacity is sufficient, much of the solar heat of the day may be retained and reemitted to the sky at night. In this situation, the incident solar heat may not impinge significantly on the interior space. If

Table 7-4. Surface reflectances.

	Reflectance	
Representative Materials and Finishes	*Solar Radiation*	*Cavity Radiation (thermal only)*
Effective for dissipation of incident radiation to the sky in warm climates		
white paint	0.71	0.11
white marble	0.54	0.05
Effective for retention of incident radiation in cool climates		
limestone	0.43	0.05
red paint	0.25	0.06
gray paint	0.25	0.05
black paint	0.03	0.05
Effective for reflection of invisible thermal radiation within a cavity space		
polished silver	0.93	0.98
polished aluminum	0.85	0.92
polished copper	0.75	0.85
aluminum paint	0.45	0.45

NOTE: Unless light-colored materials are self-cleaning or are easily and reliably cleaned at frequent intervals, much lower solar reflectances must be assumed. Generally, reflectances of 0.50 for light-colored walls and 0.10 for dark-colored walls are assumed.

there are negligible internal loads, then, the interior of a massive structure may be relatively cool on a hot day, because the interior surface radiation, indicated by the MRT, will tend to relate to the cooler temperatures of the previous night.

Massive structures will tend to produce somewhat stable interior conditions, while lightweight structures are, by contrast, much more responsive to short-term solar variations. Table 7-6 indicates time-lag requirements necessary to minimize thermal influences during peak periods. In hot, dry climates where relatively large diurnal temperature swings occur, and condensation due to high humidity levels is not a problem, massive structures are an appropriate way to maintain a cool environment.

Exterior Surface Finish

Emissivity (relating to absorption or reflection) is a related surface influence. Finishes that *reflect* and *emit* rather than *absorb* and *retain* radiation will tend to diminish the flow of energy through the shell. A portion of the incident solar radiation is in the visible spectrum (light); and white finishes are more reflective than black finishes to this portion of the spectrum. But part of the radiation involves invisible infrared and ultraviolet, and the

Table 7-5. Heat storage characteristics.

Representative Materials	Approximate Time Lag in Heat Flow through Material	
Wood		
½-in. thickness	0 hours	10 minutes
1-in. thickness	0	25
Concrete		
2-in. thickness	1 hour	5 minutes
6-in. thickness	3	50
8-in. thickness	5	5
12-in. thickness	7	50
Brick		
4-in. thickness	2 hours	20 minutes
8-in. thickness	5	20
12-in. thickness	8	20
Insulation		
2-in. thickness	0 hours	40 minutes

NOTE: For composite assemblies, the time lag is approximately additive. Add an additional ½-hour allowance for a composite assembly that involves two-layer construction. Add an additional 1-hour allowance for a composite assembly that involves three-layer construction.

Table 7-6. Time lag requirements to minimize internal thermal influences during peak periods.

Surface	Optimum Time Lag Requirements	Comments
Roof (late AM to noon peak)	10-12 hours	
East wall (AM peak)	12-17 hours	Requirement is not generally practical (too long); design to provide for little or no lag, and accept a near instantaneous load to coincide with cooler morning air temperatures
South wall (noon peak)	6-10 hours	
West wall (PM peak)	5-10 hours	Most significant wall; along with roof, this is the most useful area for massive materials
North wall	5-10 hours	Solar load is relatively insignificant in the temperate zone

reflectivity of this energy is a function more of the molecular composition and surface density than of color.

Polished metal surfaces tend to reflect significant portions of the total incident radiation. Painted surfaces tend to reflect significant quantities of light energy, but they absorb much of the radiation in the infrared range. When exposed to the sky, however, the diffuse painted surface will tend to lose significant quantities of the absorbed energy; while the metallic surfaces will retain this heat. This accounts for the prominent use of white rather than metallic exterior finishes in warm climates.

When exposed to a dark cavity, the white finish will no longer intercept significant heat by reflection and reradiation, and the metallic surface will now provide a more effective barrier to infrared heat flow. This accounts for the use of reflective metallic sheeting within insulating air spaces. (See table 7-7.)

Conductive Interaction with the External Air Mass

The previous discussion of heat storage and interception relates primarily to solar radiation that impinges on opaque assemblies. Gains and losses also

Table 7-7. Thermal resistance rates (°F-sq ft/hr)/Btu-inch of material.

Representative Materials and Assemblies	*Resistance (R)/ inch*
Aluminum, steel, copper	Negligible
Wood	
soft (fir, pine)	1.25
hard (maple, oak)	0.91
Concrete, varies with moisture content	0.08-0.11
Stucco	0.20
Gypsum or plaster board	0.90
Brick	
common	0.20
face	0.11
Glass	0.10
Insulation	
blanket and batt:	
mineral fiber	3.50
rigid boards and slabs:	
cellular glass	2.86
polyurethane	6.25
extruded polystyrene	5.00
expanded polystyrene	3.85
loose fill: cellulose	3.45
	R/Designated thickness or surface
Air spaces without reflective coatings	
¾″ width in vertical wall	0.94
¾″ width in roof	0.77
3½″ width in floor	1.22
Air spaces with reflective coatings	
¾″ width in vertical wall	2.77
¾″ width in roof	1.66
3½″ width in floor	8.17
Air surfaces	
still interior air	
ceiling	0.61
wall	0.68
floor	0.92
exterior moving air	
15 mph winter wind	0.17
1.1 mph summer wind	0.25

(continued)

Table 7-7 (continued).

Representative Materials and Assemblies	Resistance (R)/ inch
Windows (includes surface coefficients): "U"	
	Btu/hr-sq. ft.-°F
single glazing	1.11
double glazing	0.57
triple glazing	0.38

occur due to conductive transfers (through the assembly) between interior and exterior air masses.

The intensity of this conductive air flow will depend on the temperature difference between the two air masses (H = heat flow $\sim$ ($T_{outside} - T_{inside}$)); this will, of course, fluctuate with the time of day and season of the year. Heat flow will always be from the hotter air mass to the cooler air mass. On a warm day, heat gain will take place to the interior. At night or on a cold day, the interior will lose heat to the exterior.

In essence, the transmission process is analogous with the absorption and transmission of moisture by a porous material. Heat is transferred by conduction through successive layers of the enclosure until the effect is felt on the opposite surface. Some materials and layers will offer little resistance, absorbing and conducting heat rapidly. Others will exhibit a capability to reflect, store, or otherwise resist the heat flow.

Surface Configuration

Exterior surface temperatures tend to be higher than the temperature of the surrounding exterior air mass. As a result, air movement over these exposed surfaces will increase the dissipation of heat from the enclosure shell. Since this is a negative influence in cold weather, exposed surface area should be minimized in regions where cold temperatures predominate.

When surface cooling is desired, heat dissipation can be increased by utilizing techniques and forms that increase the exposed surface area. These are the use of curved surfaces such as vaults and domes, the use of uneven surfaces such as corrugated materials or uneven brick coursing, and the use of roughened surface finishes such as stucco or concrete.

Surface conductance will also increase with wind velocity; and although wind direction has a somewhat marginal effect, wind parallel to the surface will produce the highest heat transfer for a given material.

Material Resistance

Generally, materials that are good conductors of electricity are also good conductors of heat. The converse is also true.

The insulation value of a material is the resistance (R) of that material (table 7-7). The reciprocal of the resistance is the measure of positive heat flow through a material (U = Btu/sq ft-hr-°F). The thermal performance of an assembly of materials is defined in terms of the sum of the individual material resistances (R_{total}, fig. 7-5). The higher the "R" value, the greater the resistance to conductive heat flow. "R" is defined in terms of the degrees F. differential required for one Btu to be conducted through the assembly or material. (°F-sq ft-hr/Btu). Another way to describe this term is in units of time necessary for a Btu to be conducted through an assembly or material [(Hr-sq ft-°F)/Btu].

As heat flows through a homogeneous material there is a continual shift in temperature. Increasing thickness will therefore increase the resistance of the enclosure. When such increases in thickness are no longer practical or suitable, the resistance of a thicker material can be simulated by utilizing materials of higher insulation value (figs. 7-6 and 7-7).

Air Spaces

Still (motionless) air is an excellent insulator (table 7-6). When lightweight structures are desired or required, therefore, the materials in the enclosing shell should be assembled in a manner that will enclose or contain thin layers of air.

The optimum insulation value of a vertical air space in a wall is obtained at approximately ¾" width. At a width greater than this, the air begins to move in convective air currents and loses its capability to provide improved insulation values.

Although an air space is assumed to have zero wind velocity, heat is transmitted across the space by convection, radiation, and by conduction through structural members. In this regard, very significant improvement in insulation value is achieved by adding a metallic reflector coating on the cold side of the cavity space. This coating acts as a reflector of the radiant energy that is emitted across the air space from the warm side.

Estimating Conductive Heat Loss and Heat Gain

Calculations of conductive heat loss and heat gain evaluate (1) the surface area involved, (2) the transmission coefficient of the assembly (U), and (3) the design temperature differences.

For winter heat loss calculations, the difference in temperature between the interior and exterior air masses is used. The interior temperature is defined to be the comfort temperature of the occupants of the space. The design exterior temperature can be determined from climatic weather data as defined in tables found in the *ASHRAE Handbook of Fundamentals* for major weather stations. (See table 7-8.)

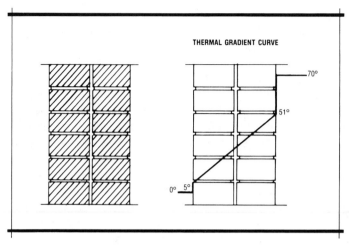

Thermal Analysis

Thermal Resistance [(°F-sq ft-hr)/(Btu)]
Assembly: 8-inch common brick wall (see table 7-7 for data)

	R
Outside surface (15 mph wind condition)	0.17
8-inch common brick	$8 \times 0.2 = 1.60$
Inside surface	0.68
	$R_{TOTAL} = 2.45$

Coefficient of Heat Transmission (Btu/hr-sq ft-°F.)
$U = 1/R_T = 1/2.45 = 0.408$ Btu/hr-sq ft-°F

Heat Loss Analysis (Conduction)
Interior Temperature: 70°F (T_I)
Exterior design temperature: 0°F (T_O)
Wall area being considered: 100 sq ft
Heat Loss $= H_C = (area) \times (U) \times (T_I - T_O)$
$= (100) \times (0.408) \times (70° - 0°)$
$= 2856$ Btu/hr loss at 0°

Heat Gain Analysis (Conduction)
Cooling Load Temperature Differential (CLTD): Table 7-9
90° design temperature, Medium daily temperature range
South wall, CLTD $= 11°$
Wall area $= 100$ sq ft
Heat gain $= H_C = (Area) \times (U) \times (CLTD)$
$H_C = (100) \times (0.408) \times (11°) = 450$ Btu/hr heat gain

Analysis of Winter Gradients through Assembly

$$\frac{0.17}{2.45} \times (70° - 0°) = 5°$$

$$\frac{1.60}{2.45} \times (70° - 0°) = 46°$$

$$\frac{0.68}{2.45} \times (70° - 0°) = \frac{19°}{70°}$$

Inside surface temperature: 51° (to be used for MRT calculations)

Figure 7-5. Basic brick wall.

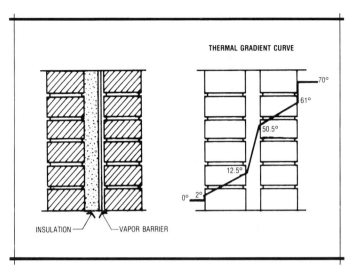

THERMAL GRADIENT CURVE

INSULATION ⎯⎯ ⎮ ⎮ ⎯VAPOR BARRIER

Thermal Analysis

Thermal Resistance [(°F-sq ft-hr)/(Btu)]
Assembly: 8-inch brick wall with 1″ cellular glass insulation (see Table 7-7 for data)

		R
Outside surface (15 mph wind condition)		0.17
4-inch common brick	$4 \times 0.2 =$	0.80
1-inch cellular glass insulation	$1 \times 2.86 =$	2.86
4-inch common brick	$4 \times 0.2 =$	0.80
Inside surface	$=$	0.68
	$R_{TOTAL} =$	5.31

NOTE: 1-inch thickness of cellular glass insulation (R = 2.86) is equal in thermal resistance to an additional 14.3-inch of common brick (R = .20 inch).

Coefficient of Heat Transmission (Btu/hr-sq ft-°F):
$U = 1/R_T = 1/5.31 = 0.188$ Btu/hr-sq ft-°F

Analysis of Winter Gradients through Assembly:

$\dfrac{0.17}{5.31} \times (70° - 0°) = 2°$ drop at outside surface

$\dfrac{0.80}{5.31} \times (70° - 0°) = 10.5°$ drop through outer brick

$\dfrac{2.86}{5.31} \times (70° - 0°) = 38°$ drop through insulation

$\dfrac{0.80}{5.31} \times (70° - 0°) = 10.5°$ drop through inner brick

$\dfrac{0.68}{5.31} \times (70° - 0°) = \underline{9°}$ drop at inside surface

$70°$ total

Inside surface temperature: 61° (to be used for MRT calculations)

Dew-point Analysis:
Interior condition: 70°, 30% RH
Dew point = 37° (fig. 7-12)
Therefore, condensation will occur within the insulation, unless a vapor barrier is placed on the warm side of the dew point.

Figure 7-6. Addition of insulation.

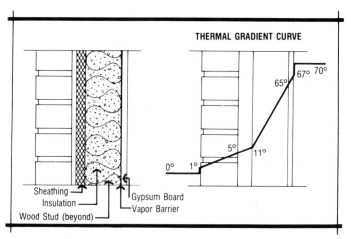

Thermal Analysis

Thermal Resistance (°F-sq ft-hr/Btu)
Assembly: Nominal 2 × 4 stud wall at 16-inch o.c. with 3½-inch mineral fiber
  batt insulation between studs and brick veneer exterior face

		R @ Insulation	R @ Wood Frame
Outside surface (15 mph wind condition)		0.17	0.17
4-inch common brick:	4 × .20 =	0.80	0.80
½-inch cellular glass sheathing	.5 × 2.86 =	1.43	1.43
3½-inch mineral fiber batt insulation	3.5 × 3.5 =	12.25	
Nominal 2 × 4 softwood stud	3.5 × 1.25 =		4.38
½-inch gypsum board	.5 × 0.90 =	0.45	0.45
Inside surface		0.68	0.68
	R_{TOTAL} =	15.78	7.91

Based upon the assumption that 15% of the wall is composed of wood framing,
the Coefficient of Heat Transmission (U) is:

$$U = 1/R_T = (0.85 \times 1/15.78) + (0.15 \times 1/7.91)$$
$$= 0.073 \text{ Btu/hr-sq ft-}°F$$

Analysis of Winter Gradients through Insulation Assembly:

$$\frac{0.17}{15.78} \times (70° - 0°) = 1° \text{ drop at outside surface}$$

$$\frac{0.80}{15.78} \times (70° - 0°) = 4° \text{ drop through brick}$$

$$\frac{1.43}{15.78} \times (70° - 0°) = 6° \text{ drop through sheathing}$$

$$\frac{12.25}{15.78} \times (70° - 0°) = 54° \text{ drop through insulation}$$

$$\frac{0.45}{15.78} \times (70° - 0°) = 2° \text{ drop through gypsum board}$$

$$\frac{0.68}{15.78} \times (70° - 0°) = \underline{3°} \text{ drop at inside surface}$$
$$70°$$

Inside surface temperature: 67° (to be used for MRT calculations)

Dew-point Analysis:
  Interior condition: 70°F., 50% RH
  Dew point = 50° (fig. 7-12)
    Therefore, condensation will occur within the insulation, unless a vapor
    barrier is placed on the warm side of the dew point.

Figure 7-7. Basic wood frame wall.

Table 7-8. Representative exterior design conditions.

City	Winter		Summer	
	°F	DD$_{65}$	°F	Range, °F
Portland, Maine	−1	7305	84	22
Boston	9	5775	88	16
New York City	15	5022	89	17
Cleveland	5	6351	88	22
Grand Rapids	5	6777	88	24
Chicago	2	6151	91	15
Minneapolis/St. Paul	−12	8060	89	22
Bismarck	−19	8992	91	27
Seattle	27	5281	82	19
San Francisco	40	3238	71	14
Salt Lake City	8	5975	95	32
Denver	1	6083	91	28
St. Louis	6	4860	94	21
Atlanta	22	3070	92	19
Miami	47	185	90	15
New Orleans	33	1392	92	16
San Antonio	30	1579	97	19
Phoenix	34	1382	107	27
Las Vegas, Nevada	28	2399	106	30
Los Angeles	40	1494	89	20

Temperature Ranges: Low <16°; Medium = 16°-25°; High >25°
Winter Design: 97.5% temperature exceeded
Summer Design: 2.5% temperature exceeded
SOURCE: ASHRAE, 1989, *Handbook of Fundamentals* & *Standard 90.1-1989.*

For estimating summer heat gains, however, *Cooling Load Temperature Differences* (CLTD, table 7-9) for opaque building envelope systems are substituted for simple outdoor–indoor temperature differences. (See sample heat loss and heat gain calculations, figure 7-5.) CLTD values represent temperature differences adjusted to compensate for both air mass temperature differences and solar radiation effects.

Because of the solar radiation effects, this heat gain load becomes a dynamic or time-dependent load. An exhaustive calculation involves thermal lag and its impact on the time of day loads. An accurate analysis includes a simulation of the weather and building program. Thus, a computer analysis is ultimately required.

Transparent building envelope systems do not generally require the analysis of thermal lag. The conductive thermal loads and solar radiation impact are considered instantaneous. Therefore, *Glass Load Factors* (GLF) (table 7-10) are used to determine the heat transfer through transparent surfaces.

Table 7-9. Cooling load temperature differences (CLTD).

	Design Temperature, °F											
	85		90			95			100		105	110
	L	M	L	M	H	L	M	H	M	H	M	H
Walls												
North	8	3	13	8	3	18	13	8	18	13	18	23
NE & NW	14	9	19	14	9	24	19	14	24	19	24	29
East & West	18	13	23	18	13	28	23	18	28	23	28	33
SE & SW	16	11	21	16	11	26	21	16	26	21	26	31
South	11	6	16	11	6	21	16	11	21	16	21	26
Roofs/Clg.	42	37	47	42	37	51	47	42	51	47	51	56

Assumes dark color surfaces and average weights.
SOURCE: ASHRAE, 1989, *Handbook of Fundamentals.*

Table 7-10. Glass Load Factors (GLF): Regular double glass (Btu/hr-sq ft) includes both conduction and solar heat gain.

	Design Temperatures, °F (Table 7-8)					
	85	90	95	100	105	110
No inside shade						
North	30	30	34	37	38	41
NE & NW	55	56	59	62	63	66
E & W	77	78	81	84	85	88
SE & SW	69	70	73	76	77	80
South	46	47	50	53	54	57
Horizontal skylight	137	138	140	143	144	147
With draperies, venetian blinds or translucent roller shades fully drawn						
North	16	16	19	22	23	26
NE & NW	29	30	32	35	36	39
E & W	40	41	44	46	47	50
SE & SW	36	37	39	42	43	46
South	24	25	28	31	31	34
Horizontal skylight	71	71	74	76	77	79

Adjust by approximately +4% for each degree of latitude above 40° to 48° north and −4% for each degree of latitude below 40° to 32° north.
SOURCE: ASHRAE, 1989, *Handbook of Fundamentals.*

An alternative method of calculation is to use the actual exterior-interior air temperature differentials (as in the winter heat-loss calculation) plus the solar radiation gain (tables 7-1 and 7-2) at a critical time of the day and year for the transparent surfaces.

CLTD calculation procedures are relatively simplified, applicable to single-family residential, some multi-family, and small commercial buildings, and for preliminary design analysis.

Infiltration

Infiltration is the accidental air flow through the enclosing shell while *ventilation* is the controlled air flow that is intended to maintain the quality of the interior air. In both cases, heat and moisture transfers occur.

With ventilation, these transfers are generally localized and the air may be subject to preconditioning before it enters the space. In contrast, infiltration is a more uncontrolled condition that involves random air changes through leaks, cracks, and other openings. Control of infiltration is primarily a problem of joint design and construction around window, door, and structural intersections (figs. 7-8 and 7-9).

The porosity of materials, however, may also be a factor in infiltration. With brick or concrete block, for example, air leakage tends to be relatively high. Application of nonporous films, such as plaster coatings or building paper, will significantly reduce, but not eliminate this leakage.

In addition to material and construction variables, the actual infiltration air volume and, therefore, heat transfer, will depend on the wind velocity and stack effects in the building.

Wind Effects

Wind will tend to increase static pressure on the outside face on one side of the enclosure shell, while the static pressure on the opposite side of the shell is reduced. As a result air will tend to flow from exterior to interior through the windward wall (or walls) and from interior to exterior through the opposite (leeward) wall.

As noted in the previous discussion of surface conductance, heat flow through the wall will also vary somewhat due to wind direction. Maximum surface conductance will occur on walls that are parallel to the direction of wind flow (fig. 7-10).

These wind effects tend to become greater as the height above ground level increases. For example, wind velocities at the upper floors of a 40-story building are approximately 1½ times the velocity at ground level. Since static pressure increases approximately as the square of the wind velocity, such an increase will have a significant effect on the infiltration rate and volume of outside air introduced into the building.

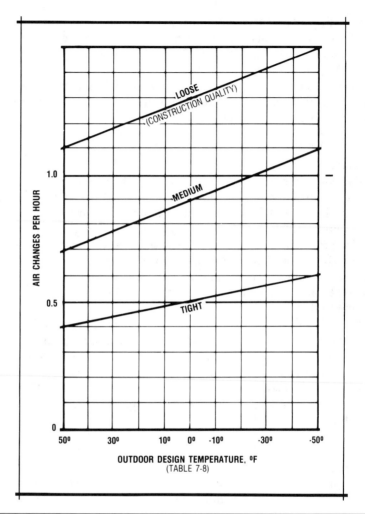

Building	Outdoor Design Temperatures, °F									
Quality	50	40	30	20	10	0	−10	−20	−30	−40
Tight	0.41	0.43	0.45	0.47	0.49	0.51	0.53	0.55	0.57	0.59
Medium	0.69	0.73	0.77	0.81	0.85	0.89	0.93	0.97	1.00	1.05
Loose	1.11	1.15	1.20	1.23	1.27	1.30	1.35	1.40	1.43	1.47

Information from Table 62, ASHRAE 89, p. 26,58

Figure 7-8. Winter infiltration air changes per hour (ACH) as function of exterior air design temperature and quality of building construction.

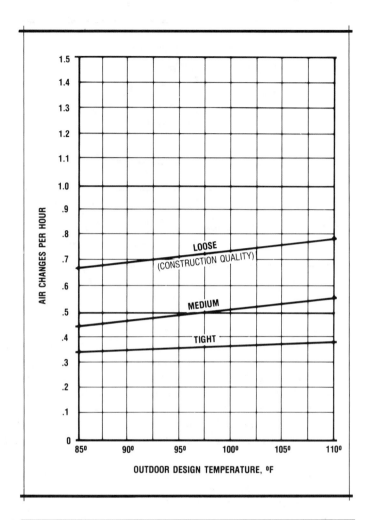

| Building | Outdoor Design Temperatures, °F | | | | | |
Quality	85	90	95	100	105	110
Tight	0.33	0.34	0.35	0.36	0.37	0.38
Medium	0.46	0.48	0.50	0.52	0.54	0.56
Loose	0.68	0.70	0.72	0.74	0.76	0.78

Information from Table 63: ASHRAE 89, p. 26,58

Figure 7-9. Summer infiltration air changes per hour (ACH) as function of exterior air design temperature and quality of building construction.

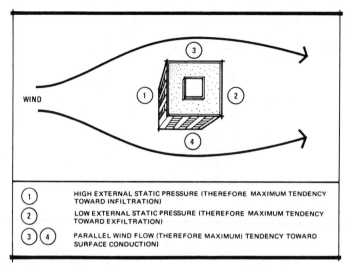

WIND

①	HIGH EXTERNAL STATIC PRESSURE (THEREFORE MAXIMUM TENDENCY TOWARD INFILTRATION)
②	LOW EXTERNAL STATIC PRESSURE (THEREFORE MAXIMUM TENDENCY TOWARD EXFILTRATION)
③ ④	PARALLEL WIND FLOW (THEREFORE MAXIMUM) TENDENCY TOWARD SURFACE CONDUCTION)

Figure 7-10. Wind effects.

Stack Effects

Because of the natural tendency for warm air to rise, significant vertical variations in interior static pressure are induced, and these tend to produce a pronounced *chimney effect* in tall buildings. This will induce infiltration on all walls at the lower floor levels, while inducing exfiltration (outflow) at the upper floors. Near the middle floor levels, there is a *neutral zone* at which there is little pressure gradient for air flow either way except for the previously discussed wind effects. Figure 7-11 illustrates the influence of the stack effect.

This *stack effect* is minimized by sealing all openings between floors such as stairways, elevator shafts, and mechanical and electrical shafts. In most cases, fire safety requirements will mandate this sealing of vertical openings.

Large, open vertical spaces, such as atria, provide ample opportunity for stack effects to occur. As a thermal system, this type of space is acceptable because human occupancy exists at the floor level and at observation points around the periphery. Hot air at the top of the space does not influence the occupants' comfort.

Atria natural convective air flow can also be used as a return or exhaust air system. An opportunity to use the space as a smoke shaft with exhaust at the top and human observation points around the periphery to locate a fire and its by-products such as smoke also exists.

Pressurization

Since infiltration of outdoor air will carry dust and contaminants as well as inducing locally variable heat gains and losses, conscious effort is often

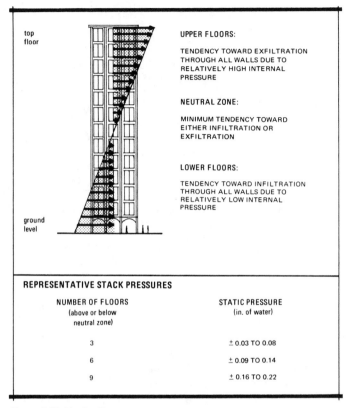

Figure 7-11. Stack effects.

made to minimize this effect by pressurizing the building interior. This is accomplished by reducing the ventilation exhaust capacity relative to the ventilation intake or fresh air supply capacity, thereby causing a rise in interior static pressure.

This applied force will tend to induce exfiltration through all cracks, openings, and porous materials. For this reason, it must be of sufficient pressure to balance or counteract the infiltration that results from wind or stack effects.

Vapor/Air Barriers

Condensation of water vapor will occur within wall, roof, or floor assemblies at the temperature plane where seeping air becomes saturated. This temperature plane is defined by the dew point temperature of the air mass on the warmer side of the enclosure shell. (See figures 7-6 and 7-7 and the related dew point analysis.) Figure 7-12 illustrates the determination of the dew point from the psychrometric chart.

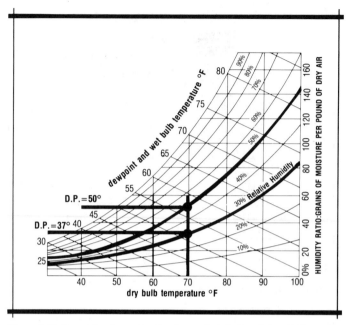

Figure 7-12. Dewpoint determination for analysis in figs. 7-6 and 7-7.

Since moisture can cause physical damage and deterioration within the assembly, it is necessary to minimize this effect. One method involves the reduction of the moisture level on the warm side of the barrier. Relative humidity levels must remain high enough, however, to maintain human health and comfort standards. To prevent vapor and air transmission and subsequent condensation within the assembly, therefore, an impervious air/vapor barrier should be placed between the theoretical dew point plane defined at the maximum design temperature condition and the moisture-laden warmer air mass.

The nearer the vapor barrier is to the moisture source, the greater will be the moisture limit that can be tolerated in the warm air mass. Ideally, then, the vapor barrier should be placed near the inner surface in cold climates and near the outer surface in hot-humid regions. Since the greatest temperature and absolute humidity differences between air masses occur during the winter months in north temperate regions, the vapor barrier is generally placed near the inner surface in these climate zones.

Internal Heat Gains

Not too many years ago, builders of human enclosures were dealing with situations where the heat generated by the occupants was the only source of internal heat other than man-made heating devices. The major problem in

Table 7-11. Heat gains from occupants of conditioned spaces.

Degree of Activity	*Total Heat*	*Sensible Heat*	*Latent Heat*
Seated, very light work	400	245	155
Moderately active office work	450	250	200
Standing, light work; walking	450	250	200
Light bench work	750	275	475
Walking 3 mph; light machine work	1000	375	625
Heavy machine work; lifting	1600	635	965
Athletics	1800	710	1090

Based on 75° room temperature and on normal percentage of men, women, and children. Gain from women is 85% and gain from children is 75% of that for an adult male.
SOURCE: ASHRAE, 1989, *Handbook of Fundamentals.*

thermal design was to control heat excesses or deficiencies related to weather or climatic conditions.

This situation has changed. Today, many building types are thermally dominated by the influence of internal heat. This is true for building types such as office buildings, manufacturing facilities, and hospitals where computers and lighting systems contribute large amounts of electrically generated heat during normal hours of use. In addition, new standards of building shell thermal insulation have increased the thermal containment of this internally generated heat within the building shell. Many buildings are capable of maintaining comfort conditions to comparable low exterior temperatures during hours of occupied conditions. Even in cold climates the thermal design problem becomes one of providing cooling to balance the excess internal heat.

Because of this phenomena, it is increasingly important that internal heat gains be evaluated to determine their impact on traditional concepts of orientation and building envelope design. Methods of control and generation of internal heat are significant aspects of thermal design.

Heat Gains Associated with Occupancy and Ventilation

Human occupancy requirements produce significant quantities of internal heat. Part of this heat is associated with occupant heat itself—that is heat generated by the human metabolic processes (table 7-11). For individuals, this heat tends to be somewhat minor when compared with other conduction and radiation transfers. But this heat becomes a significant factor when the density of occupancy increases. This category of heat gain is particularly significant for enclosures intended for congregational and assembly functions.

A second category of occupancy heat gain is associated with the ventilation air that is introduced for hygienic purposes (table 3-6). Each occupant requires an adequate supply of fresh air. Since exterior air replaces interior air, this process may introduce adverse temperature and humidity conditions into the interior environment.

Heat Gains Associated with Equipment

Equipment installed for specific needs and functions is the most recent category of greatly increased internal heat contribution. For example, the increasingly prevalent use of electronic equipment such as computers, communication, and information systems is adding significantly to internal heat.

Estimates of a generally distributed equipment load typically begin at one watt per square foot of floor space, but with the introduction of computer systems to a space, this load can be estimated up to 15 watts per square foot. Special concentrated loads should be analyzed and designed separately. (See table 7-12.)

In analyzing the internal thermal balance, some of this heat is in a form and location that permits partial isolation and control. For example, some large machines and computer systems are located in remote, isolated, and separate areas. A portion of this heat can therefore be harnessed through manipulation of air circulation or fluid flow by means of techniques that conduct considerable quantities of heat away before it can penetrate sufficiently to become a direct influence on occupancy comfort.

Table 7-12. Preliminary estimates for heat gain from equipment.

Function	Watts/Sq Ft
Assembly	0.25
Office	0.75
Retail	0.25
Warehouse	0.10
School	0.50
Hotel or motel	0.25
Restaurant	0.10
Health	1.00
Multifamily	0.75
Specialized loads: Verify each program	
Computer room	15 to 50
Laboratories	5 to 20
Office with computers	up to 6.0

SOURCE: ASHRAE/IES 90.1-1989, "Average Receptacle Power Densities."

Heat Gains Associated with Electric Lighting

The use of electric lighting produces another significant heat gain. This applies both in situations where the lighting is intended to implement visual performance in work spaces and where spatial systems provide environmental lighting for orientation and circulation, assembly, or relaxation. (See table 7-13.) In both cases, the heat generated by these systems may become a significant influence in the thermal environment. Improved efficiency lighting sources, ballasts, luminaires, and appropriate daylighting systems can reduce this electric energy load and the corresponding cooling loads caused by the electric loads.

In initially evaluating the thermal influence of the electric lighting system, the designer must estimate the action of the electric lamp as a source of heat as well as a source of light. Heat is emitted at the rate of 3.413 Btu/watt-hr of energy consumed.

The availability of this internal energy may be a decisive consideration in thermal system design. If this energy remains randomly distributed and

Table 7-13. Preliminary estimates for heat gain from lighting systems.

Function	Watts/Sq Ft
Auditorium	1.6
Classroom/lecture hall	2.0
Food service	
fast food/cafeteria	1.3
leisure dining	2.5
Library, reading area	1.9
General lobby	1.0
Office	
reading, typing, filing	1.8
drafting	2.6
Church, synagogue, chapel	2.5
Dormitory bedroom	1.1
Hospital	
patient room	1.4
operating room	7.0
Hotel guest room	1.4
Museum, exhibition	1.9
Theater, performing arts	1.5
Merchandising general	3.1
Mall concourse	1.4

SOURCE: Based on Standard ASHRAE/IES 90.1-1989, "Unit Lighting Power Allowance."

uncontrolled, it can become a somewhat chaotic thermal influence. But, if this energy can be effectively harnessed, it can be utilized or exhausted as necessary to respond to demands imposed by the outdoor environment. Of particular interest, then, is the somewhat isolated location of many lighting devices and their character as heat traps.

The lamp is generally placed in an enclosure of some type, such as a luminaire or ceiling cavity, and this begins to alter the distribution of heat. When the lamp is first turned on, the ballast heat (if there is a ballast), the heat of the lamp itself, and the heat absorbed by the air immediately surrounding the lamp is all trapped within the fixture or cavity. This heat may not remain in this location for a long period of time; but for a short period of time it is confined in an enclosure somewhat remote from the immediate environment of the occupant.

Furthermore, some of the emitted light is absorbed as heat by the luminaire itself. The invisible infrared and ultraviolet energies follow a similar path, depending on the material and finishes of the luminaire itself. Again, part of the energy is emitted immediately into the room, while the remainder is absorbed in the fixture housing or cavity.

By utilizing the luminaire or cavity as a heat trap, a substantial quantity of lamp and fixture heat can be recovered by moving return air through the luminaire. The return air system can easily control heat within the duct system, and up to 80 percent of the heat can be captured before it enters the occupied space. An additional benefit is that the luminaire system itself is cooled so that it no longer acts as a radiant heating panel and allows the lamps to operate at temperatures where greater lumen efficiency is achieved.

Effective removal and control of lighting heat by exhaust air requires a low velocity air path that moves past the lamp. This circulation immediately past the light source is an important factor in the efficient collection of heat.

Because exhaust velocities are low, then, regularly spaced vents or slots should be placed as near to the lamp as possible in order to channel a maximum quantity of this low velocity air through the immediate environment of the lamp itself.

At the same time, it is equally important that this air be channeled to move along the metal fixture surfaces, to pick up radiant energy that has been intercepted and absorbed by these surfaces. The regular spacing of air slots (rather than using a single slot) will help this aspect by facilitating more uniform air motion through the system.

Quantitative Estimates of Thermal Loads

The building envelope heat transfer issues and the internal building program heat discussed above can be analyzed through quantitative techniques. The calculation techniques are based upon the analysis of the instantaneous heat flows occurring in the design environment correlated with the climatic conditions of the natural environment within which the man-made environment exists. A complete analysis requires a great deal of detail and information and can become a tedious and time-consuming activity. Computer techniques are almost mandatory for any large building and any building for which a great deal of accuracy is desired.

There are, however, manual techniques from which computer techniques have been developed and which in themselves allow the individual designer to investigate the system requirements and the annual energy consumption expected within the program parameters. These techniques are particularly appropriate for small buildings and as preliminary approximate sizing of systems and energy consumption for all types of buildings.

Tables 7-14 and 7-15 give an outline of the manual procedures which are typically utilized to determine thermal loads. Figures 7-13 and 7-14 are examples of calculations for a small building.

Annual Energy Costs

Heat loss and heat gain calculations are designed to determine the thermal loads at a point-in-time design condition or for extreme conditions. The determined loads are then the net size of the heating and cooling equipment so as to be able to balance the thermal loads at those extreme conditions.

These calculations do not tell us the operating energy costs. Operating energy costs are based upon average weather conditions as opposed to the extreme design weather conditions utilized to size the equipment. In order to determine the annual fuel costs, we can either (1) utilize computer

Table 7-14. Estimation of enclosure heat losses.

Component	Formulas	Total	References
Opaque walls	$(U_w) \times (Area_w) =$	Btu/hr-°F	Fig. 7-5, 6, 7
Glass	$(U_{gl}) \times (Area_{gl}) =$	But/hr-°F	Table 7-7
Roof	$(U_r) \times (Area_r) =$	Btu/hr-°F	Table 7-7
Infiltration	$(ACH) \times Volume_{cf} \times 0.018$		
or	or		
Ventilation	$(CFM) \times 1.08 =$	Btu/hr-°F	Fig. 7-8
Floor slab	$.5 \times$ perimeter length	Btu/hr-°F	
Building Load Coefficient (BLC) =		Btu/hr-°F	

Building heat loss = $H = (BLC) \times (T_i - T_0)$
  H = Design heat loss or furnace/boiler size in Btu/hr
  T_i = Design inside temperature
  T_0 = Design outside temperature

Fan size = cfm = $(Btu/hr)/(1.08 \times (T_{SUPPLY} - T_{RETURN}))$

Table 7-15. Estimation of enclosure heat gains.

Component	Formulas	Total	References
Transmission gains:			
opaque walls	$(U_W) \times (Area_W) \times CLTD =$	Btu/hr	Table 7-9
roof	$(U_R) \times (Area_R) \times CLTD =$	Btu/hr	Table 7-9
glass	$(GLF) \times (Area_{GL}) =$	Btu/hr	Table 7-10
Infiltration			
	$ACH \times Volume \times 0.018 \times (T_0 - T_i) =$	Btu/hr	Fig. 7-9
Internal Gains:			
occupants	# occupants × sensible		
	heat gain/occupant =	Btu/hr	Table 7-11
lighting	(watts/sq ft) ×		
	(sq ft) × 3.4 Btu/watt =	Btu/hr	Table 7-13
equipment	(watts/sq ft) ×		
	(sq ft) × 3.4 Btu/watt =	Btu/hr	Table 7-12

Total Sensible Heat Gain (SHG) = Btu/hr
This load is used to size the air-handling equipment by the formula:
$$cfm = \frac{SHG}{1.17 \times (T_{SUPPLY} - T_{RETURN})}$$

The size of the refrigeration equipment must also include the latent heat load. For preliminary estimates, simply add 30% of the sensible heat gain as latent heat gain to the sensible heat gain. This sum is the preliminary size of the refrigeration equipment.

A more accurate method of determining the latent heat gain is through use of the enthalpy values on the Psychrometric Chart.

techniques which simulate the normal yearly weather conditions for a particular location and climate or (2) utilize manual techniques which translate the equipment size to an estimate of what the fuel costs would be for that equipment in that particular environment.

The manual technique that is utilized for calculating annual fuel consumption and costs for heating is based upon the concept of degree days. A *degree day* is the difference between the exterior temperature at which no heat is required to heat the interior of a building and the daily exterior mean temperature. Generally, this exterior temperature is assumed to be 65° in the heating season unless there are large amounts of internal heat and/or solar heat. For example, if the mean temperature for a 24-hour day is 40°F, there would be $65° - 40° = 25$ degree days accumulated for that particular 24-hour cycle. The weather bureau documents the degree-days accumulated over the course of the annual heating cycle. Each city then has an average annual degree day quantity that represents the climatic heating requirements of the location. See table 7-8 for typical values.

The net annual heating energy requirement is determined by the formula:

Btu/year = (Btu/hr-°F) × (degree days/yr) × (24 hours/day)

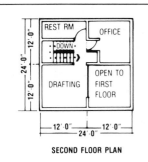

SECOND FLOOR PLAN

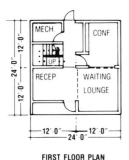

OFFICE BUILDING

FIRST FLOOR PLAN

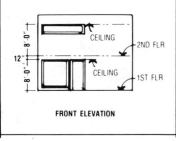

FRONT ELEVATION

Location: Grand Rapids, Michigan
    Winter design temperature: 5°F
    Interior design temperature: 70°F

Exterior wall area: 1700 sq ft
20% exterior wall area = glass
    20% × 1700 sq ft = 340 sq ft double glazing
        (U = 0.57 Btu/hr-sq ft-°F)
    80% × 1700 sq ft = 1360 sq ft wood frame wall,
        figure 7-7,
        (U = 0.073 Btu/hr-sq ft-°F)
Roof: 8″ mineral fiber insulation
R = 8 × 3.5 = 28°F-sq ft-hr/Btu
U = 1/R = 1/28 = .036 Btu/hr-sq ft-°F

Heat Loss

System	Calculation			Btu/hr-°F
	(U)		(Area)	
Wall	0.073	×	1360 sq ft =	99
Glass	0.57	×	340 sq ft =	194
Roof	0.036	×	576 sq ft =	21
Infiltration	0.75 ACH × 9000CF × 0.018 =			122
Floor edge	.5 Btu/hr-°F-LF × 96 LF		=	48
	Building Loss Coefficient (BLC) =			484
				Btu/hr-°F

Design transmission heat loss = net furnace size:

= Building Loss Coefficient × $(T_i - T_0)$ =
(484 Btu/hr-°F) × (70° − 5°) = 31,460 Btu/hr
  T_i-interior design temperature
  T_0-exterior design temperature

If we assume an 80% efficient gas furnace and gas costs at $6.00/1000 cubic feet, the annual estimated energy costs are:

Btu/yr = (Btu/hr-°F) × degree days/year × 24 hr/day
    = (484) × 6777 × 24 = 80,000,000 Btu/yr

$$\$/yr = \frac{(Btu/yr) \times (\$/unit\ fuel)}{(Btu/unit\ fuel) \times System\ efficiency}$$

$$= \frac{80,000,000 \times \$6.00/1000}{1000 \times 0.80}$$

= $600.00/year

Figure 7-13. Small office building: Example heat loss calculation.

247

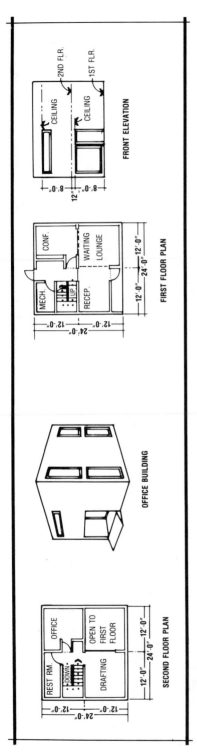

Location: Grand Rapids, Michigan
Summer design temperature: 85°
Interior design temperature: 75°

Building envelope conditions are the same as for the heat loss calculation.

Heat Gain

System	Calculation							Btu/hr
	(U)		(Area)		(CLTD)		=	75 Btu/hr
Wall					3°F			
North	0.073	×	340 sq ft	×	13	=		323
East	0.073	×	340	×	6	=		150
South	0.073	×	340	×	13	=		323
West	0.073	×	340	×	37	=		767
Roof	0.036	×	576	×		=		
	(GLF)		(Area)					
Glass	30 Btu/hr-sq ft	×	85 sq ft			=		2550
North	77	×	85			=		6545
East	46	×	85			=		3910
South	77	×	85			=		6545
West								
Infiltration								
0.46 ACH × 9000CF × 0.018 × (85°−75°)							=	745

System	Calculation		Btu/hr
Internal Gains			
People	10 people × 230 Btu/person	=	2300
Lights	2.5 w/sq ft × 1000 sq ft × 3.4	=	8500
Equipment	1 w/sq ft × 1000 sq ft × 3.4	=	3400
	Total sensible heat gain (SHG)	=	36133 Btu/hr

Latent Heat Gain (LHG)
Estimate 30% of SHG
30% × 36133 = 10840 Btu/hr = LHG

Equipment Size = SHG + LHG
36133 Btu/hr + 10840 Btu/hr = 46973 Btu/hr
@ 12000 Btu/ton, refrigeration equipment size =

$$\frac{46973 \text{ Btu/hr}}{12000 \text{ Btu/ton}} = 4 \text{ tons}$$

Figure 7-14. Small office building: Example cooling load calculation.

The annual costs for heating energy are determined by the formula:

$$\$/yr = \frac{(Btu/yr) \times (\$/unit\ fuel)}{(Btu/unit\ fuel) \times (system\ efficiency)}$$

See table 7-16 for typical values for different fuels. Economic costs can be determined based upon current local fuel costs.

Balance Point Temperature

Large buildings with significant amounts of internal heat and smaller buildings which are well insulated and, thus, do not lose heat quickly, are not exclusively controlled in the thermal environment by the exterior conditions. In recent times, internal loads have increased significantly through increased use of electronic equipment such as computers and communication systems and increased reliance on electric lighting systems. In this condition, the building program is a primary determinant of the thermal environment, and cooling becomes a more important issue in thermal design.

Similarly, the increased emphasis on thermal insulation in the building envelope has tended to hold internally generated heat and solar heat for longer periods of time within the building skin. Again, the result of this emphasis is increased cooling loads and decreased heating loads.

The analysis of this phenomena can be defined within the concept of the *balance point temperature (BPT)*.

$$BPT = T_i - \frac{IHG}{BLC}$$

where:

BPT = Temperature at which the internal heat balances with the heat flow through the building envelope.
T_i = Inside design temperature (°F)
IHG = Internal heat gain at a point in time (Btu/hr)
BLC = Building load coefficient (Btu/hr-°F)

Table 7-16. Heating systems fuel comparison.

Fuel	System Efficiencies	Typical Btu/Unit Fuel
Gas	.5-.95	1000 Btu/1000 C.F.
Oil	.5-.80	140,000 Btu/gallon
Electricity	1.00	3,413 Btu/kW
Wood	.1-.5	20-30,000,000 Btu/cord

In the example office building problem of figures 7-13 and 7-14, at a condition of full occupancy and use of full equpment and lights, the IHG is:

People:	2300 Btu/hr
Lights:	8500
Equipment:	3400
Total IHG:	14200 Btu/hr

$$BPT = 70° - \frac{(14200 \text{ Btu/hr})}{(484 \text{ Btu/hr-°F})} = 70° - 30° = 40°$$

At 40° external temperature, the internal heat generated at these full-occupancy conditions exactly balances the heat flow to the outside through the building envelope. No heating or cooling is needed to balance the thermal loads to satisfy human comfort. Above 40° the building must be cooled in order to maintain the 70° interior comfort temperature. Below 40° the building must be heated to maintain the 70° comfort temperature. In calculating annual energy, adjustment must be made in the number of degree hours to compensate for the reduced heating requirements.

It must be recognized that the full-occupancy state is an extreme condition which will be achieved in only a limited number of hours. A full analysis of the thermal operating costs requires a comparison of the building program loads and schedule with the climatic weather conditions and the number of hours of temperature ranges. The condition of full external envelope heating load with no internal gain must still be provided through the sizing of the heating system.

Manipulation of the BPT can be achieved through changing the insulation levels of the exterior envelope, varying the quantities and types of glass, and isolating in time and space the internal heat generated. Passive solar heat is similar to the IHG in its impact on the BPT.

BALANCING SYSTEMS

In the final analysis, it is difficult, if not impossible, and many times not economically justified, to provide thermal comfort conditions only through the design of the building envelope. For example, buildings in the temperate zone are exposed to an extreme range of climatic influences, ranging from extreme cold, to extreme hot and humid conditions. The optimum enclosure response to the cold conditions is a compact, well-insulated building, while the optimum response to the hot-humid conditions is an open structure that will facilitate the penetration of cooling breezes.

This requirement for multiple solutions is difficult to develop in a single building form that will satisfy all of the weather variations that will occur throughout the year. Generally, the best solution appears to be the one that satisfies the intermediate conditions with as many hours as possible of the typical weather year while leaving the extreme conditions for a mechanical thermal balancing system. For this reason, a variety of energy-balancing devices has been developed to compensate for the periodic discrepancies that will inevitably occur when attempting to optimize the response of the building to varying climatic influences.

Heat-generating Devices

Several categories of systems are utilized to provide heat in the interior of a building. The most obvious solution would appear to be the combustion process in the traditional sense of a fireplace. But modern technology has developed noncombustion processes as well.

Noncombustion Systems

Among *noncombustion* types is one that will be discussed later under the heading of *cooling devices.* This is a combination heating-cooling system that is capable of providing heat during periods of heat loss. The system is the *heat pump,* which utilizes condenser heat for heating and evaporative cooling for cooling needs.

A second type of noncombustion system is *electric resistance heating.* Electric current run through a resistor generates heat which can be utilized to heat a space directly.

A third type of noncombustion process is the utilization of *solar energy.* Solar energy can be utilized in a passive manner wherein the building construction components serve as the heating system. South-facing glass in the northern hemisphere (north-facing glass in the southern hemisphere) provides a transparent opening for the solar radiation to penetrate to the interior of the building. The radiation continues until it strikes an opaque dark-colored mass, where it is absorbed. This dark colored mass is typically an integral component of the building structure such as a wall or the floor. An auxiliary space such as a sunspace or a greenhouse can also serve as a solar heat collector and the collected heat transferred as desired to the habitable environments.

The absorbed heat is transferred to the ambient air of the interior environment by conduction/convection or radiation. The emitted radiation is a long-wave radiation as opposed to the shorter wavelength of the solar radiation. This change in wavelength contributes to the amount of heat trapped in the interior environment, as glass allows transmittance of the short-wave incoming solar radiation but resists transfer of the long-wave interior emitted radiation. Therefore, by controlling the transmission of radiation and conductive heat, glass contributes to the addition of solar generated heat to the interior environment. Figure 7-15 describes the typical passive solar heating system concepts.

In contrast to the passive solar heating systems which use no auxiliary energy to operate the system, the *active* solar heating concept utilizes

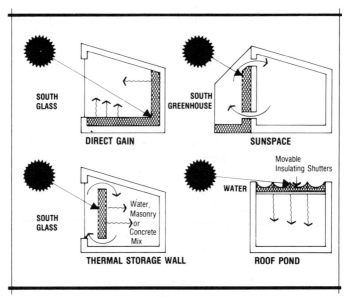

Figure 7-15. Passive solar heating systems.

electrical energy to operate a fan or pump to transfer the collected solar heat (see fig. 7-16). Dark-colored surfaces with a separated glass layer, or *solar collectors,* are mounted on the building facing as close to directly south as possible (northern hemisphere) with a tilt to the horizontal ground plane at approximately the angle of the building's latitude. These collecting surfaces absorb the solar radiation. Water or air is passed over the collecting surface to transfer the heat to a remote storage mass. The storage mass is water, stones, or a latent heat storage material. The stored heat is then utilized by an active distribution system which transfers the heat to the environment as desired. Often this type of system is augmented by a more typical heating system which serves to provide the necessary heat during non-sunny periods.

Combustion-type Systems

Combustion-type devices are based upon the traditional fireplace concept. Fireplaces utilize natural fuels such as wood and peat. Because of (1) the inherent inefficiency of an open fireplace, (2) the requirement that fireplaces be located in the spaces they serve, resulting in multiple installations with corresponding chimneys, and (3) the human labor requirements of placing the fuel periodically in the fireplace, central *fossil fuel* heating devices were developed. These centralized heating systems made a significant contribution to modern architecture by freeing the designer from the constraints of several heat sources located in the occupied spaces and

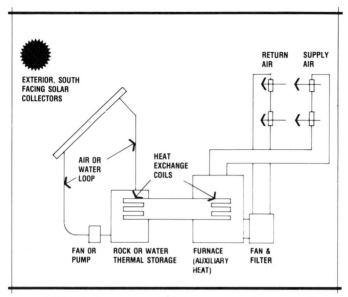

Figure 7-16. Active solar system: Solar heat augments heat supplied by the furnace.

allowing the building envelope to develop a form with minimal constraints on the distance from the heat source or on the amounts of fenestration. Unfortunately, it also provided the opportunity to ignore local and regional components of the environment and to consume natural resources with little awareness that they were being consumed. These systems utilize natural gas, oil, or coal to heat water or air which is then distributed throughout the building.

Selection of Fuel and System

A transportation system for the fuel must be in place. For gas this requires a pipeline; oil and coal require vehicular transport, while electricity requires a power and transformer distribution network. Space requirements within the building for generating equipment, for fuel storage, for chimneys and for distribution devices is another consideration. The ultimate goal of the system is to provide for the thermal comfort and stimulation of the occupants of the environment. A comparative economic evaluation must be made of the potential systems based upon the costs of system installation, energy, and operation. The environmental costs are often hidden in the fuel rates and costs. The designer must be cognizant of the current and future environmental consequences of the decision.

In the systems comparison chart (table 7-17), fuel costs tend to be low for combustion-type processes as compared to noncombustion or electrical

Table 7-17. Comparison of building heating systems.

System	Comfort		Environmental Impacts		Fuel
	Control	Consistency, Quality	Source	Exhaust, By-products	Transport
Noncombustion					
Heat pump	Thermostat	Low temp.	Elect. gen plant	Gen plant by-products	Electricity
Electric resistance	Thermostat	Very good	Elect. gen plant	Gen plant by-products	Electricity
Active solar	Thermostat	Poor—need aux source	Sun	None	None
Passive solar	Building mass	Function of climate/design	Sun	None	None
Combustion					
Wood	Damper	Poor	Tree removal	High CO_2	Vehicle
Coal	Thermostat	Fair	Earth distortion	High CO_2	Vehicle
Natural gas	Thermostat	Very good	Limited	Low CO_2	Pipeline
Oil	Thermostat	Very good	Limited	Low CO_2	Vehicle

| | Fuel (Continued) | | Equipment | | | | Distribution | |
System	Storage	Cost	Generation	Chimney	Cost	Medium	Outlet	Fan/Pump
Noncombustion								
Heat pump	None	High	Package + heat sink	No	High	Air	Register	Fan
Electric resistance	None (or mass)	High	Off-site	No	Low	Electricity	Radiant device	None
Active solar	Water or rocks	Low (fan or pump)	Collectors	No	High	Air	Register	Fan
Passive solar	Building mass	Free	South fenestration	No	High	Radiation + natural conv.	None	None
Combustion								
Wood	Required	Low	Combustion chamber	Yes	Medium	Radiation + air	Radiation + register	Natural or fan
Coal	Required	Low	Combustion chamber	Yes	Medium	Air or water	Register	Fan or pump
Natural gas	None	Low	Combustion chamber	Yes	Medium	Air or water	Register or convector	Fan or pump
Oil	Tank	Low	Combustion chamber	Yes	Medium	Air or water	Register or convector	Fan or pump

systems while equipment needs are relatively expensive. Because of the extensive equipment requirements of combustion systems, initial equipment costs and maintenance costs tend to be high. Electrical system production occurs off-site, and the equipment production and maintenance costs, as well as the environmental costs, are reflected in the relatively high fuel costs for electrical energy.

Consequently, most electrical systems present less extensive generating requirements. Chimneys and fuel storage are eliminated; space for equipment is often reduced; and initial equipment costs may be lower. Energy costs will be higher.

Each project must be evaluated as a unique event. The solution and corresponding environmental and economic factors must be carefully studied.

The Combustion Process and Distribution of the Heating Medium

As fuels are burned in combustion systems, oxygen is required to support the combustion. Since air contains only about 20 percent oxygen, very significant volumes of air must be brought to the immediate proximity of the burner where the fuel is located. Inadequate flow of combustion air results in uneconomic use of fuel, as well as the production of carbon wastes—both soot and excessive quantities of carbon monoxide. For these reasons, optimum quantities of air and proper flow through the burner are decisive requirements.

The chimney is a basic element in controlling the flow of combustion air and gases. It actually performs two functions: inducing a flow of air (draft) through the combustion chamber and carrying off the wastes (smoke and flue gases). In combustion systems, the combustion process itself defines one route-of-flow through the heat generation system (figure 7-17). Heated flue gases tend to rise through the chimney by natural or forced convection. This process not only evacuates undesirable gases, but it also induces a flow of fresh air into the combustion chamber to replace the exhausting gases. This constant flow of fresh air supplies the oxygen required to sustain the combustion process.

The second route-of-flow is completely separate and relates to the heat distribution medium. This medium is warm air in a furnace or hot water in a boiler. It is initially collected around the perimeter of the combustion chamber, where it receives heat by conduction through the metal heat exchanger chamber enclosure. As this medium is heated, normal gravity convection will begin to induce a natural circulation, with the heated air or water rising through a duct or pipe system.

After the circulating medum discharges its heat to an occupied space, the cooled medium will tend to seek a lower level. A system of return ducts or pipes is provided for this; this completes the cycle back to the perimeter of the combustion chamber, where the reheating process begins again. This

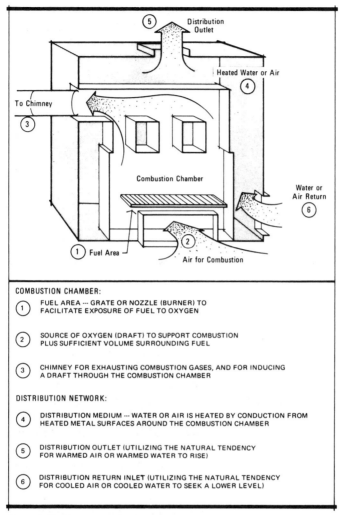

Figure 7-17. Heat generation (combustion).

distribution process can utilize natural gravity flow, as described, or fans and pumps can be introduced to achieve a more variable or precise rate of flow (fig. 7-18).

Cooling Devices

Natural Cooling

When heat removal is required to bring building conditions into the comfort zone, the natural heat sink of the sky, the atmosphere, and the

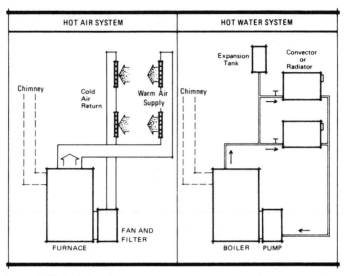

Figure 7-18. Basic heating systems.

earth's mass are available as cooling resources. Passive cooling techniques include cross-ventilation and stack ventilation, which rely on movement of the air and consequent displacement of the warm air by cooler air from the atmospheric heat sink; mass heat storage, which relies on both radiation to the sky and air movement for heat transfer to the atmosphere; and earth coupling, which transfers heat to the earth heat sink. Also, evaporative cooling is available in dry climates, whereby the sensible air temperature is reduced by the evaporation of water into the atmosphere. This increases the relative humidity and thus is not appropriate or effective in warm, moist climates. Comfort conditions may be exceeded for air moisture content, and the air will not have a large capacity to absorb more moisture.

Ventilation techniques require the availability of cooler outside air to reduce the interior air temperature. Cross-ventilation relies on wind velocity to provide the motive force for allowing this cooler air to penetrate the interior. There must be both inlets and outlets for this air. Generally, the design approach is to provide window or louver openings in the direction of the prevailing summer winds and outlet openings at the opposite end of the enclosed space.

Stack ventilation relies on the natural tendency of warm air to rise and be replaced by cooler air. If a clear vertical dimension of 8 to 10 feet is achieved in a space with openings at the top and bottom of the vertical shaft, the natural convection of warm air will occur. As with the cross-ventilation technique, the outlet opening must be at least as large as the inlet opening, and preferably larger than the inlet opening.

Mass cooling relies on the principle that mass requires heat to increase its temperature, and the more mass available, the greater amount of heat that can be stored per degree temperature rise. When utilized in conjunction with ventilation techniques, it is an effective way to store and remove internal heat. A typical system is designed to collect and store heat during occupied hours and then to cause cooler outside air to come in contact with the mass storage surfaces during the nonoccupied night hours. Also, by exposing the mass to the night sky, radiant heat loss can be utilized.

Mechanical Cooling

When natural techniques are not feasible or desirable, excessive heat and humidity that accumulates in the interior space can be eliminated by refrigeration. Early cooling systems accomplished cooling by the process of fusion, such as the change of state of ice to water (solid to liquid). Heat from the overheated space was absorbed at a rate of 144 Btu/pound of water (the heat of fusion for water) in the conversion of ice to water.

The heat absorbed by one ton of ice melting over a 24-hour period is 144 Btu/pound × 2000 pounds or 288,000 Btu per day. This equals 12,000 Btu per hour. Hence, *one ton* of refrigeration capacity is 12,000 Btu/hr, the unit utilized to describe the size of a refrigeration system.

Contemporary refrigeration systems utilize the process of the heat of evaporation to accomplish cooling action. Again, there is a change of state that causes the refrigerant to absorb heat from the interior air mass. If water is the refrigerant, for example, the purpose of the system may be to create conditions in which the water will evaporate. When this is done, the *heat of evaporation* will absorb approximately 1000 Btu from the interior air mass in the conversion of each pound of water to vapor.

Since the useful refrigeration effect depends on the quantity of heat absorbed in the change of state, the evaporation process is more efficient than the fusion process. Furthermore, the ability to transport, store, and process liquids is generally more effective than it is for solid refrigerants.

In order to complete the refrigeration cycle, the refrigerant must be reconverted to the original state. For evaporative processes, this is generally accomplished by condensing the refrigerant vapor and expelling the stored heat to a *heat sink*. This heat sink may be the outdoor air mass, a water well or lake, the earth, or a cooling tower. In any case, the greater the temperature difference between the condenser heat and the heat sink, the more effective is the system in dissipating its heat and returning the refrigerant to the original state.

Compression Devices for Cooling

Cooling by mechanical compression is shown diagrammatically in figure 7-19. This action involves a refrigerant, typically a *chloroflurocarbon (CFC),*

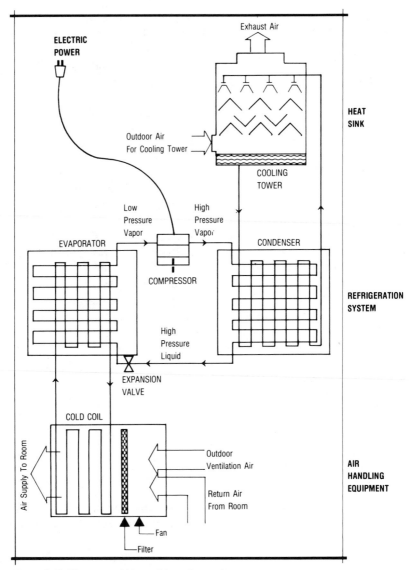

Figure 7-19. The compression refrigeration cycle.

transported in a closed system under varying pressures in order to control the capability of the refrigerant to take on or give off heat.

The environmental consequences of a leaking CFC refrigerant, or a CFC refrigerant allowed to escape once it is no longer being used, has become a major concern. Because of their great stability, CFCs eventually diffuse into the stratosphere where they break down and release chlorine, which combines with the ozone molecules and diminishes the protective ozone layer.

The ozone layer serves to prevent excess, harmful solar ultraviolet radiation from reaching the earth's surface. In the lower atmosphere, the CFC molecules absorb infrared radiation and contribute to the greenhouse effect and the warming of the earth. (See earlier discussion concerning the "Solar Environment.")

Consequently, a great deal of research and investigation is occurring to develop new types of refrigerants which will not pose such grave environmental consequences. The implementation of these new types of refrigerants will require time and expenditures for conversion to new processes and the new equipment involved.

The actions involved in the compressive refrigeration process are:

1. A high pressure liquid flows from the *condenser* to the *evaporator* through an expansion valve so as to relieve the pressure.

2. Cooling action occurs due to the evaporation of the refrigerant in the low pressure chamber of the evaporator. The heat of evaporation of this process is drawn from the interior air mass of the occupied environment.

3. The resulting low pressure refrigerant vapor from the evaporator passes to the *compressor,* where mechanical compression of the gas increases the pressure to the saturation condition.

4. Under compression, the refrigerant vapor returns to the condenser and condensation occurs as the saturated vapor gives off its heat. The stored heat is ejected to outdoor air or to cooling water. The cooling water then passes into a well, to a cooling tower, or to an air-cooled condenser.

5. The resultant high pressure liquid flows to the evaporator, and the cycle repeats itself.

The system becomes a heat pump when it is designed to be reversed and, in effect, the evaporator and condenser change places. In this condition, the evaporator draws heat from a heat sink, such as the outside air, a natural body of water, heat stored from interior heat, or solar heat. The condenser supplies this heat to the interior environment. A system which combines a series of heat pumps in a water loop provides a technique of shifting cooling load rejected heat to areas that require heating.

In cold climates the exterior chamber which draws heat from a heat sink can freeze. This not only reduces the efficiency of the system but also may require periodic defrosting of the chamber.

The ratio of the output energy to the input energy is called the *coefficient of performance (COP)*. This value can range from less than 1.0 to 4.0 or higher. The critical factor is the temperature of the heat sink and the design of the heat pump system to draw that heat from the sink.

Absorption Devices for Cooling

In the compression refrigeration cycle, the input energy is electricity driving the compressor. This is an expensive system to operate because the vapor undergoes a large change in specific volume. If it were possible to raise the pressure of the refrigerant without altering its volume, the required work energy could be reduced.

The *absorption refrigeration cycle* accomplishes this with absorption of the refrigerant vapor by a liquid (such as water) (fig. 7-20). Although a very significant reduction in work by moving parts is accomplished by this method, heat input must be increased several times over that required for the mechanical compression cycle. For this reason, an inexpensive source of heat is required in order to make the absorption cycle attractive.

As in the mechanical compression system, the useful refrigeration effect of an absorption system depends on the change of state (enthalpy) within the evaporator. However, several additional components are required to perform the function of the compressor/condenser combination. The function of these components is as follows:

1. An *absorber* containing a salt solution absorbs water vapor from the evaporator. This action increases the tendency of the water in the evaporator to evaporate, thereby cooling the bulk of the water in this component. A

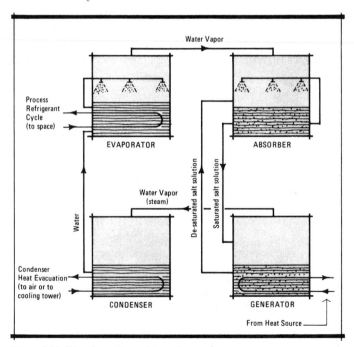

Figure 7-20. The absorption refrigeration cycle.

heat-exchange coil transfers this cooled effect to the air handling unit or the occupied environment.

2. A *generator* is introduced into the system to maintain the salt solution in the absorber at a proper concentration. The weakened salt solution that has become saturated with water is pumped from the absorber to the generator. The excess water is boiled off in the generator. This boiling action requires the heat input for the system. This heat can be waste heat from a source such as an electric generating plant or a waste disposal plant, solar generated heat, or the product of a fossil fueled source such as a coal or gas heating system. The restored strong salt solution is then pumped back to the absorber to repeat the absorption process.

3. The boiled water or vapor is pumped to a condenser where it condenses back to its liquid form. The heat that is rejected goes to a heat sink such as a cooling tower. The restored water is then pumped back to the evaporator to repeat the entire process again.

Throughout the process the only moving parts are the circulating pumps.

Thermal Storage for Cooling

A major portion of the energy costs for the cooling of buildings is due to the electrical demand charge which utility companies assess to account for the capital costs of building an electric generating facility. An electric power generating plant accounts for the major expenditures in providing electric energy. Because the utility company must have adequate generating facilities for the peak demand of the yearly, weekly, and daily cycles, the size of the plants is determined by the peak demands and not the average demands. Consequently, the generating capacity is oversized for the majority of the hours.

In order, then, to recapture some of their capital expenditures, which do not produce in the way of sold electricity, the utility companies assess a demand charge. This additional charge is based upon the user's peak demand during a billing period and is in units of kilowatts (as opposed to kilowatt-hours).

The rate is correlated with the time of day and year according to the times the utility might expect to provide its peak services. Typically, this occurs in the late afternoon of a mid-summer day and is determined by large space cooling loads.

In order to avoid or minimize this demand charge and decrease the required peak electric generating capacity, many building owners, encouraged by utility companies, are including thermal storage systems within the building's HVAC system. This allows the cooling system to be run during off-peak hours, typically at night and during weekends, minimizing the hours of operation during high demand and, consequently, reducing costs. Cooling is generated within a storage tank and stored in water, ice, or eutectic salts. When the peak demand occurs at a later hour, the stored cooling can be utilized by the building's cooling system.

Not only does this type of system reduce demand and operating costs, it can also reduce the required size of the cooling equipment, which initial savings can be utilized to pay for the storage system. This type of technology provides operating savings while at the same time providing ecological and resource conservation benefits.

Air-Handling Equipment

Ultimately, the heating and cooling produced for the comfort of the occupant of the environment must be transferred to the ambient air of the environment. The exception is in the design of radiant systems that interact with the body surface to provide comfort. This system requires an additional air-handling system in order to provide a supply of fresh treated air for the health and comfort of the occupants.

The heating and cooling can be transferred to the occupied environment by water, where a heat exchange occurs between the water and the ambient air. This has the advantage of minimizing the duct requirements and saving volume within the building. However, a minimum fresh air requirement exists which can be satisfied only by some type of air-handling system. Also, because comfort in hot summer conditions and high density occupied conditions involves the removal of moisture, or condensation, in the air, some form of condensation removal must occur in a cooling system. This is difficult to accomplish within the occupied space. Therefore, most large buildings which supply cooling will have some kind of central air-handling system.

In small buildings the air-handling capability is located in the furnace. If a hot water heating system is utilized, fresh air requirements can be provided through infiltration.

In larger buildings with central air-handling equipment, a connection must be made between the cool-producing evaporator and the air-handling equipment. The air-handling equipment typically consists of a fan, filters, and ducts for return air, supply air, and fresh air intake. Connection is made with the cooling and heating sources by heat exchange coils, typically water lines, which transfer the heat (see fig. 7-19).

In the cooling process, chilled refrigerant is pumped from the evaporator to the air-handling equipment. A fan circulates warm room return air, outdoor air, or a mixture of both over a coil. This action extracts heat from the warm air, thereby cooling and dehumidifying the air and preparing it for circulation to the conditioned space.

The extracted heat is absorbed by the process refrigerant, which moves back to the evaporator and is cooled as the cycle begins anew.

Humidity Control

The easiest manner in which to remove moisture from a given air mass is to reduce its temperature, so as to bring the air mass to saturation and cause the necessary moisture to condense out of the air. Temperature reduction and dehumidification are both accomplished by the evaporator coil. Air to be conditioned is drawn over the coil, where it is cooled to near saturation and water vapor is condensed and drained away. The air is then mixed or reheated as necessary to provide the desired environmental supply condition.

Filters

Special conditions or building programs may require localized filtering of ambient air. But all air must be filtered at some point. Therefore, filters are typically located in the central air-handling equipment. They process both recirculated and outdoor intake air. Types of filtering devices are:

1. *Dry filters* such as cloth, felt, cellulose, or wire screens. Regular maintenance is required for cleanliness or replacement.

2. *Spray washers* not only clean the air but can control the humidity as well by adding moisture to the supply air.

3. *Electric precipitators* are based on the principle of attraction of static charges. Dust particles and other particle impurities can be withdrawn from the air by attraction to an electrically charged surface.

Demand Zones

From the central heating and cooling systems and the air-handling equipment the conditioned air or water must be distributed throughout the building. The purpose of a distribution system is to distribute the conditioning medium in the best possible manner to allow the occupants to have thermal comfort throughout the building. This leads to the concept of *demand zones.*

Because of the variable nature of solar and internal loads, interior heating and cooling demands at different locations throughout a building may be quite variable at any point in time. Demand zones are the division of the building into thermal zones with individual control in response to unique thermal loads.

Types of loads to consider are:

• Solar loads which vary with time of day and season and the amount of glass occurring on each elevation.

• Internal program loads which vary with the function of each space and the time of day and week.

• Exterior envelope conduction loads which vary with the amount of surface area in contact with the exterior environment (walls and roof) and the thermal insulation resistance of the envelope.

• Functional and orientation factors which influence the infiltration of the outside air due to wind or building entry/exit locations.

Typically, a demand zone serves a 1000–2000 square foot area. Ideally, the optimum control would be to make each room or definable area an individual demand zone with a thermostat to control system performance in each area. A residence is typically a single zone. For larger buildings the problem is dividing the building into groups of rooms or areas that are constantly exposed to similar thermal influences, with each zone serving unique thermal requirements.

The exterior areas of a building are a most obvious beginning point to zone a building. When wall and roof openings are prevalent, solar and wind loads may vary considerably. East, south, and west exterior areas are alternatively exposed to varying daily solar intensities, while the north exterior area is exposed to minimal solar influence. Solar loads also produce seasonal variations. For example, south rooms experience heat gain on a clear winter day while adjacent north rooms are experiencing heat loss. The conditions can change quickly. When a cloud obscures the sun, the south wall will shift from heat gain to heat loss conditions similar to the north wall.

Figure 7-21 illustrates that a good zoning organization is to define exterior zones in the southeast, southwest, northwest, and northeast corners. These exterior zones may extend as little or as far from the exterior of the building as the distribution system will feasibly allow and as the designer wishes to define a functional zone.

Also, a vertical separation should be defined to account for the varying influences of the roof and ground or entry points to the building and the *stack effect* prevalent in tall buildings.

This leaves the internal zones which, for many large buildings, account for the majority of the building area. These internal zones are not influenced by exterior weather conditions and typically produce only heat. The problem is typically one of providing only cooling and adequate fresh air. The division into zones is one of function and program control.

Varying tenants and schedules and different functions, such as computer rooms with very high heat generation, auditoriums with time-varying heat generation and fresh air requirements, and open office spaces with rather consistent demands under full occupancy conditions are issues to be addressed.

Distribution Systems

The distribution system must provide a heat transfer cycle in which the internal air mass is periodically and methodically moved past the critical

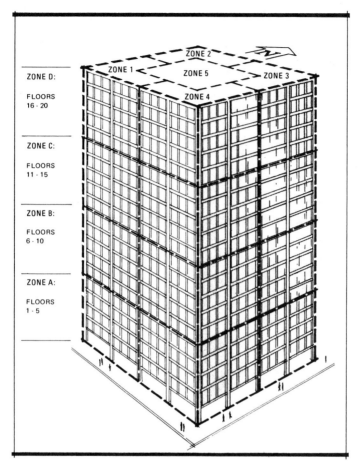

Figure 7-21. Vertical and horizontal demand zoning.

room surfaces where heat transfer takes place. The system must then exhaust the immediate occupied area by moving the air flow to a more remote mechanical room location where it can be processed and finally reintroduced to the occupied area, where the cycle is repeated.

The distribution can be by air, water, or a combination of the two. Some air distribution must be available to satisfy fresh air requirements. However, an all-air system requires large volumes of duct space within the building and the operation of large fans which consume large amounts of energy. Because of this, many large buildings have a combination of air and water systems.

Thermal Potential in the Demand Zone

The potential of supply air for modifying the temperature of an interior air mass is dependent on two factors: the volume-rate of the supply and return air and the temperature difference between the supply air and the interior room air mass (or return air). The quantity of air in an air distribution system required to provide the thermal requirements of a space is determined by the formula:

$$CFM = \frac{H_a}{1.08 \times (T_s - T_r)}$$

where:

CFM = volume in cubic feet/minute of supply air required to provide the necessary thermal balancing energy
H_a = the thermal load at design conditions in Btu/hour for cooling or heating, whichever gives the largest volume of supply air
T_s = temperature of the supply air, °F
T_r = temperature of the room air at design conditions (this is the return air temperature), °F
1.08 = (density of air) × (specific heat of air) × (60 minutes/hr)
= (.075lb/CF) × (.241Btu/lb-°F) × (60)

The manipulation of either or both the CFM and $(T_s - T_r)$ quantities is the basic function of the design of the air-handling system.

For a radiant hot-water heating system, the formula is:

$$GPM = \frac{H_h}{20° \times 60 \text{ minutes/hr} \times 8\#/\text{gallon} \times 1 \text{ Btu}/\#}$$

$$= \frac{H_h}{9600}$$

where:

GPM = Gallons per minute
H_h = Design heat loss in Btu/hr-°F
$20°$ = Temperature difference of hot water at the radiant convector
$1 \text{ Btu}/\#$ = Specific heat of water

Variable Air Volume System (VAV)

The variable air volume system has become one of the more popular systems today because of its economy of operational, maintenance, and

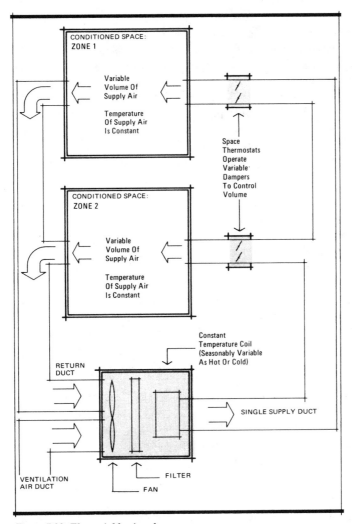

Figure 7-22. The variable air volume system.

installation costs. It operates on the principle of manipulation of the air volume with a constant air temperature (fig. 7-22). It is a system limited in flexibility because the air can only be cooled or heated. Consequently, if a zone requires heating while another zone requires cooling, an auxiliary system must be provided.

Comfort is achieved with a variable rate of air change, which must vary within prescribed limits. The air change must be high enough to avoid stagnation and contamination, and yet not so high as to introduce drafts and cold spots. Generally, a limit of 20 air changes per hour for high velocity systems applies as a maximum air change rate.

Because the system cannot provide simultaneous heating and cooling, auxiliary systems such as reheat and radiant baseboard convectors are often added to the system. In this way, the VAV system can provide minimum cooled fresh air while the auxiliary system provides the necessary heat to different zones.

In zones requiring only cooling, such as interior zones, the VAV can provide the thermal needs in a responsive manner to the interior functions. Varying heating and cooling demands which occur in response to periodic external climatic changes—particularly those that occur during the fall and spring seasons—are difficult to respond to adequately with the VAV system.

Constant Air Volume System

A system which responds to the external climatic changes by providing simultaneous parallel supply air flows of hot and cold conditioned air and then mixing the two air supplies at the demand zone is able to provide significant comfort and performance. Periodic zone demand requirements are met by varying the mix; this is accomplished by thermostatically controlled *mixing boxes*. The thermostatically controlled variable, then, is supply air temperature.

While this concept is effective in providing quick response to varying demand conditions that occur simultaneously within a building, there is an inherent inefficiency in the system. The process of reheating previously cooled air, or the inverse process of cooling previously heated air, is wasting energy resources. Consequently, although a popular system in years past, it is today a seldom used system in new buildings.

Reheat System

Because of the thermal response limitations of the VAV system and the energy resource inefficiencies of the constant volume/variable temperature system, the reheat system as a modification of the VAV system has become more popular. In this system, a VAV central distribution air-handling unit and main duct is created with thermostatically controlled reheat coils located in the duct at each zone that may periodically require heat (fig. 7-23).

The central VAV system has only to provide cooled air. The reheat coils provide a warming of the air as required. In this way a minimum amount of fresh, recirculated air can always be supplied to each zone. The system is basically operating on air volume control with some modification by temperature control. It is a compromise to provide improved comfort with a minimum cost in energy.

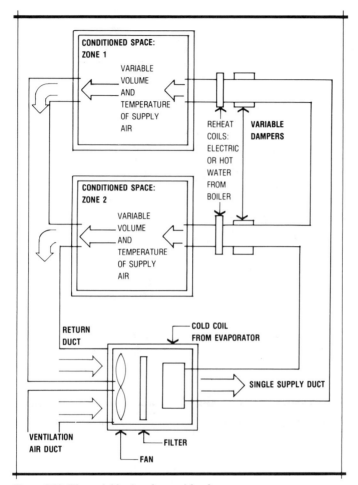

Figure 7-23. The variable air volume with reheat system.

Water System with Fan Coils

The four-pipe water system utilizes parallel hot and cold water as the distribution medium (fig. 7-24). Because water is much more compact and efficient than air as a medium for heat transfer, the space savings over an all-air system are often very significant. Large supply and return air ducts are replaced by much smaller piping.

This water system can provide quick response to the variable demand conditions that occur simultaneously in different building zones.

The fan coil system consists of a terminal fan unit located in or adjacent to the served zone which provides air handling and treatment in each individual

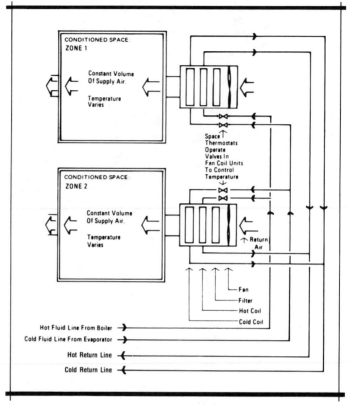

Figure 7-24. The four-pipe water system (with individual fan coils).

zone. Outside air can be brought in at each individual unit or from a central point.

Radiant Systems

Demand zone control can be achieved by means of chilled water supplied to radiant ceiling panels and heated water supplied to radiant perimeter convectors. These water systems modify the conditions so as to achieve thermal balance with a constant supply of air distributed from a central point.

These radiant systems have the advantages of providing thermal comfort through radiation, which is the most effective technique, and the ability to be located where the thermal loads occur. Maintenance is required within the zones being served and, thus, may provide a disruption to space usuage.

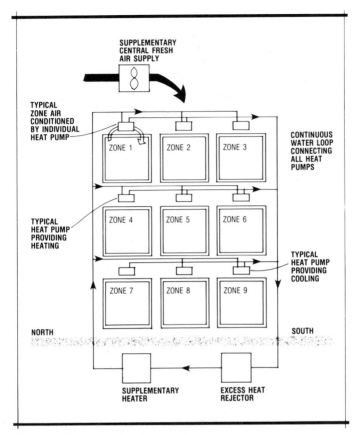

Figure 7-25. The water loop heat pump system.

Water Loop Heat Pump Systems

Individual heat pumps located in each thermal zone and connected by a continuous water loop provide a system which has good individual zone thermal control and the ability to shift thermal energies from one zone to another (fig. 7-25). In cold winter seasons, excess internal heat is removed from internal zones by the cooling cycles of the interior heat pumps and transported to the water loop, thus making the heat available for use by the exterior zones. When there is not enough internal heat produced to maintain the required temperatures, an auxiliary boiler or other heat source provides the necessary heating requirement to the water loop.

In warm summer months, all heat pumps act in the cooling cycle, and the removed heat is transported via the water loop to a heat rejector such as a cooling tower. In the moderate seasons, this system has the ability to shift quickly from the heating to cooling mode and vice-versa, depending on the climatic conditions and solar orientation. Again, transfer of the heat is readily facilitated and auxiliary heat supply and rejection is available.

The system requires that individual zone air supplies be transported over the heating/cooling coils of the individual heat pumps and that fresh air be supplied through an auxiliary air handling system. Maintenance and quality control of the individual heat pumps is the biggest limitation of the system. Access for service and adequate room air distribution are issues the designer must address.

Thermal Distribution within the Occupied Space

The ultimate objective of the heating, ventilating, and air-conditioning system is to provide for the thermal comfort of the occupants of the various spaces. Obviously, this thermal action must occur within the space itself. The components and methods of delivering this thermal energy to the space is an area of great concern for the designer.

The system must be capable of adding or extracting heat at critical surfaces and at the location of the occupants of the space. This is generally done by processing the entire interior air mass, which becomes the major vehicle for manipulation of the thermal environment.

As mentioned previously under the discussion of the thermal comfort of individuals, the body uses three means to achieve thermal comfort. They are conduction–convection, radiation, and evaporation. The systems response is through control of (1) the ambient air temperature, (2) the relative humidity or air moisture content, (3) the surface temperatures as defined by the mean radiant temperature, and (4) the movement of the air as experienced by the occupants, or *draft*.

The technology available to achieve this control is the air distribution system with the potential addition of a radiant system.

Air Distribution Methods
Natural Convection

Localized volumes of stagnant air will cause isolated extreme temperature and humidity conditions within the interior air mass. Similarly, concentrations of heat gain or heat loss can cause localized discomfort in a space that is otherwise described to have average room conditions well within the comfort zone. For these reasons, it is necessary to study air paths in the space carefully to ensure that air passes through or induces proper movement in each portion of the space.

Generally, this involves a recognition of the natural (gravity) tendency of cool air to seek a lower level and of warm air to rise. As figure 7-26 indicates,

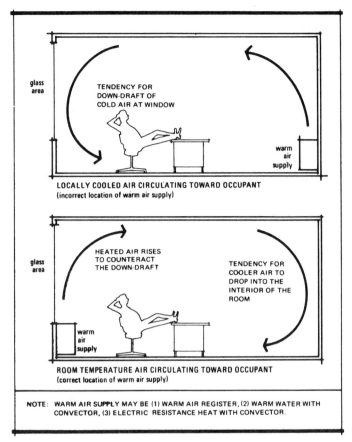

Figure 7-26. Natural gravity convection.

local cooling and warming currents are created by the placement of the thermal sources and the diffusion of the supplied air with the ambient air mass. Such convective influences are controlled, in part, by the location, type, and quantity of the air supply and return devices and, in this case, by the location of the heating element within the space.

Forced Convection

The projection and directional capabilities of forced air systems can be utilized to modify and extend the simple flow patterns induced by natural gravity convection. In this regard, the supply has two basic characteristics that determine its performance. They are the motive force and the thermal character, primarily temperature.

Duct velocities can be virtually any quantity desired. The restrictions occur due to noise created by the air friction and the energy consumed by increased velocities. Supply air temperatures can be up to 160°F in the heating season. In the cooling season temperature differentials of up to 20° below the room temperature or the return air temperature are workable. The limitation in the cooling season occurs due to discomfort created for the occupants by excessive temperature differentials and humidity factors, which cause condensation on the ducts.

The greater the air velocity and the greater the temperature differential, the smaller will be the duct size. If we ignore friction, the duct size can be calculated simply as:

$$A_d = (\text{Volume})/(\text{Velocity})$$

where:

A_d = Cross section area of the duct in square feet

Volume = Volume of air required to provide design heat loss or heat gain in cfm

Velocity = Velocity of air movement as supplied by the design fan in feet per minute.

Friction increases as the square of the increase in velocity. This imposes an increased energy load in the fan power required to provide the velocity of the air and in the increased heat added by the fan energy to the duct cooling load.

Care must be taken to avoid high velocity air streams within the occupied zone of the room, generally defined to be the volume below the 6-foot level. This limitation tends to lead to the use of ceiling or high wall supply outlets. The object of air diffusion is to create the proper combination of temperature, humidity, and air motion in the occupied zone (fig. 7-27). Discomfort can arise due to excessive room air temperature variations, excessive air motion (draft), failure to deliver or distribute air according to the load requirements at different locations, and overly rapid fluctuation of room temperature.

Draft is defined to be *any localized feeling of coolness or warmth of any portion of the body due to both air movement and air temperature, with humidity and radiation considered constant.* Every 15 fpm increase in velocity over 30 fpm is equal to a one degree drop in temperature.

Air supply outlet types are (1) grilles or registers, (2) slot diffusers, (3) ceiling diffusers, and (4) perforated ceiling panels. Diffusers are supply air outlets that control the planes and directions of the emitted air. Typically, a diffuser is a ceiling mounted outlet that directs the air horizontally across a ceiling plane. A register has no such directional control but is simply an air supply outlet grille with volume control such as a damper.

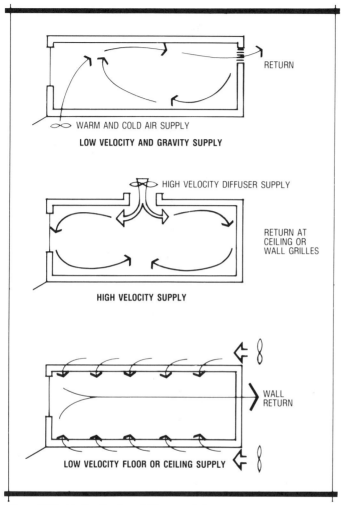

Figure 7-27. Air distribution within occupied spaces.

Low velocity air flow may be introduced into the room directly through the ceiling or floor with a large number of outlets. This requires large supply ducts. Individual air supply and control elements are available to be located at individual work stations for personal control and comfort.

High velocity air must be introduced through diffusers that develop a horizontal or near-horizontal flow across the ceiling. In this way, the supply air is permitted to diffuse and interact with the upper level strata before it moves down to the occupied zone at reduced velocity and with a reduced temperature gradient.

A typical air jet which is blown horizontally into an open area with no temperature difference between the supply and room air will tend to expand approximately 15 degrees on either side of the axis.

When supply air temperature is below the temperature of the air in the room (cooling influence), density differences will deflect the trajectory downward due to the effects of gravity. Conversely, the introduction of warmer supply air will tend to deflect the trajectory upward. In either case, the *throw* or horizontal projection distance will vary with the discharge velocity of the air jet. Velocity may also affect the angular spread of the cone of air described above. These phenomena can be used to estimate the approximate pattern of air movement from a given diffuser location.

Condensation Control

Air paths also exert an evaporating action that is important in controlling condensation.

Ventilation in concealed structural spaces such as wall, ceiling, and floor cavities is needed to facilitate the removal of water vapor that may pass from occupied areas to these concealed locations. The location of vapor barriers to prevent this moisture penetration is discussed previously. Without adequate ventilation, vapor that condenses in the colder concealed space can cause water damage and, in some cases, damage due to freeze or frost conditions.

Warm air flow up the inside face of a window tends to warm the glass, thus reducing the tendency toward condensation in cold weather. In order for this flow of air to be most effective, however, it must be constant, not intermittent. Therefore, it cannot be controlled by thermostats for the air supplied to the main space. Constant output supply systems may therefore be required in these perimeter locations.

Air Processing Limitations

When the glass area increases or the ability of the building shell to resist heat exchange diminishes in the design process, or when the program internal heat increases, the cooling thermal load inside the building increases. However, there are limitations in the ability of the conventional air-handling system to respond to these increased loads. This limit generally falls in the range of 60 to 75 Btu/sq ft of floor area.

This increased cooling load can be neutralized by increasing the rate of air change. High velocity supply air can be introduced and increased in volume up to approximately twenty air changes per hour. When this volume limit is approached and exceeded, there may be drafts and cold spots in the occupied zone.

Alternatively, the increased cooling load can be neutralized by introducing colder air and thus increasing the design temperature differential. However, as the input air is cooled below the dew point in the room, moisture will begin to condense on the ducts and metal diffuser surfaces. This condition establishes a limit of approximately 20° as the maximum temperature differential between the cooled input air and the return air in the room.

Supplemental panel cooling techniques offer one possibility for extending the thermal system capacity. But, in general, it becomes advisable and economically necessary to evaluate techniques for isolating and removing waste heat more effectively and efficiently, to harness waste heat and minimize the impingement of this excess load within the occupied space itself.

Minimizing Heat in the Occupied Zone

Substantial quantities of heat can be isolated locally. Heat can be trapped within the lighting system; solar energy can be captured or reflected by solar screen devices and solar collection systems; large quantity heat-producing equipment such as computers and communication devices can be controlled by local air-handling systems, and so forth. This principle reduces the significant quantities of heat which would otherwise have to be handled by the room air-handling system. The heat is isolated separate from the occupants of the space.

The isolated heat can now be manipulated in a number of ways. It can be rejected, stored, and/or distributed to places and at times when it will help reduce the required heating load.

The precise economic value of this isolation principle will depend on the concentration of energy involved (watts per square foot, glass area, orientation) and the feasibility of providing the best human environment without reducing its quality by the thermal isolation techniques. The principle is significant because it allows the introduction of both natural and electronic heat-producing elements in the design, without exceeding the thermal capacity of conventional systems. The effective limit is thus extended beyond the previously noted 60 to 75 Btu/sq ft. It also means there is potential for meaningful reduction in energy resource consumption.

Controlled Lighting Heat

The thermal performance of a general lighting system will demonstrate the potential of heat isolation.

Figures 7-28 and 7-29 compare conventional and luminaire exhaust systems with respect to the thermal loads calculated in fig. 7-14. Without the thermal control of air return through the luminaire, all of the lighting heat is imposed on the space as a load for the cooling system. But with the luminaire

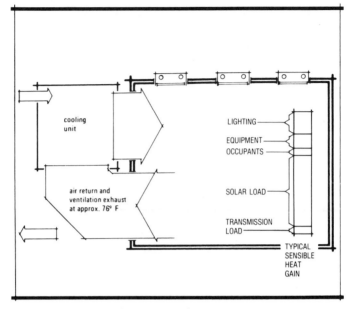

Figure 7-28. Cooling cycle conventional air return.

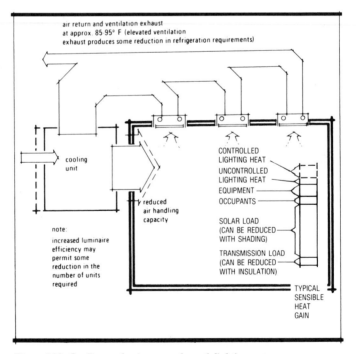

Figure 7-29. Cooling cycle air return through lighting system.

air return a large portion of the lighting heat is captured before it enters the space.

The luminaire return air system operates as follows. Room air is drawn by negative pressure through the luminaire face, passed over the lamps, and exhausted into the return air duct or a return air ceiling plenum. With this action, a substantial portion of the lighting heat is removed before it can enter the occupied space. The warmed return air can then be distributed in a number of different ways, depending on the thermal demands of the building at any given time. In cooling load conditions, the options are:

• The return air can be partially or totally rejected to the outside in moderate weather when outdoor conditions provide an abundant supply of quality replacement air.
• Return air can be retained in the building, to be tempered and regulated through proportional mixing with cooler outdoor air or with refrigerated air as required.
• The entire quantity of return air can be recycled through the refrigeration system and then redistributed.

In heating load conditions (fig. 7-30), the options are:

• The warmed return air can be redistributed to the demand zones that require heating.
• The returned air can be tempered and regulated through proportional mixing with cooler outside air.
• The return air heat can be extracted by passing it through a heat exchanger, where it can be used to temper the outside ventilation air.
• The return air can transfer the captured heat to a heat storage system so that the heat can be used during periods of higher heat demand such as nonoccupied, nonlighted hours.

Figure 7-30 illustrates the principle of internal heat balancing the building envelope heat loss. The previous discussion concerning balance point temperature demonstrates how the thermal energy balance is calculated.

Radiant Systems

The immediate thermal environment can be adjusted by manipulation of the mean radiant temperature by two principal techniques: panel radiation and concentrated radiation. Also, convection units can transfer heat to the ambient air by the radiation process in conjunction with air convection.

Panel Systems

Ceiling and floor surfaces can be utilized as supplementary or primary panel radiation devices. Walls can also be utilized, but are generally of

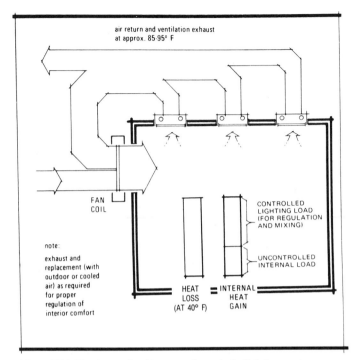

Figure 7-30. Heating cycle: Air return through the lighting system. Internal heat gain balances envelope heat loss. See discussion on balance point temperature.

limited value because of the varying and inconsistent distances from a wall panel to the occupant, and because of the variable area available for use in this way, limited by window and door openings and structure.

Probably more than other form of thermal energy distribution, panel systems must conform to the specific characteristics of the interior-exterior thermal exchange. Pipe coils should be spaced closer together near window areas, with wider spacing in the more interior areas. Furthermore, effective insulation should be positioned to minimize the inefficiencies associated with losses to the ground, to the exterior air, or unoccupied spaces.

This method is limited by the maximum values for floor and ceiling temperatures described in chapter 3. As a general rule, the capacity of a floor panel should be limited to approximately 10-20 Btu/sq ft of surface. For ceilings, the heat capacity limit is approximately 70 Btu/sq ft of surface. When the room heat loss exceeds this limitation, other means of primary or supplementary energy distribution must be utilized. The principal disadvantage of this technique is the fact that control is limited to temperature only. Such factors as humidity and atmospheric contaminants are not affected by this process.

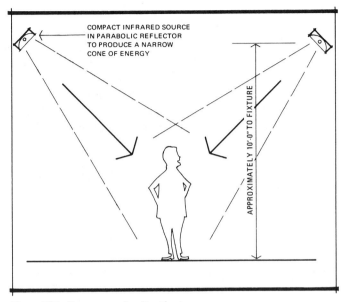

Figure 7-31. Concentrated radiant heat.

Table 7-18. Representative limited-zone radiation for interiors.

Temperature of Indoor Air (°F)	Supplementary Incident watts/sq ft (Required to Compensate for the Deficient Indoor Air Temperatures)
60°	10–25
50°	20–35
40°	40–55

NOTE: Lower intensity in above ranges applies for nondraft interior areas, where occupants are separated from cold walls. Higher intensity in above ranges applies for drafty interior areas, where occupants are exposed to cold wall conditions nearby.

NOTE: Wattages shown apply for normally clothed interior occupants (when outdoor air temperature is 0°).

Concentrated Thermal Radiation

For areas or situations where the frequency of air change precludes an attempt to process the environmental air mass itself, high intensity sources of infrared energy can be effectively controlled and concentrated for limited zone heating (fig. 7-31). Typical areas are garages, uninsulated warehouses, semiprotected waiting shelters, and building entrances. Table 7-18 indicates the representative limited zone radiation requirements for various indoor temperatures.

Essentially, these systems involve compact high intensity energy sources such as electric resistance heaters and quartz lamps that can be placed at the focal point of a specular parabolic reflector. In this way, narrow or broad directional cones of energy are emitted. To be effective, these systems should be designed to maximize surface exposure within the beam (figure 7-30).

This technique is similar in principle to the warming effect of the sun on calm days in winter when the air temperature is relatively low. These concentrating energy sources, then, can be used to heat semiexposed and localized areas independent of the environmental temperatures involved. The person or object to be heated must be effectively screened from the wind. These sources can also be used where quick warm-up response is required.

Convectors

A radiant device can be utilized to heat the ambient air in a space and not the occupants directly. The earliest example of such a device is the cast-iron radiator. Hot water or steam is circulated through these rather massive devices and allowed to give off heat both by radiation to the occupants of the space and by convection. Cold room air enters at the bottom of the unit and, as it is warmed, rises vertically through the radiator to return to the room as warmed air.

A convector is a device that encloses the radiator so as to limit the heat transfer to the circulating air only. In this way, the system becomes much more efficient, as most of the energy goes to the occupied space environment.

Convectors can be (1) a box or a surface-mounted device with damper control, (2) a finned radiator with fins attached to the circulating pipe to increase the air contact for convection, or (3) a baseboard radiation unit. The latter two types of devices are typically linear elements which blanket a wall surface with warmed air.

Convectors generally supplement local and time demands and are used in conjunction with primary air-handling systems.

Reduction of Radiant Effects

There is often a radiant heat effect associated with the electric light beam in a space, particularly when incandescent systems are involved. During the cooling season, this radiant energy is difficult to eliminate with the usual air circulation methods because the transfer of heat is directly between two objects or surfaces and does not involve room air. The following techniques are useful for reducing such radiant heat effects:

• *System efficiency:* Utilize lighting systems with high efficacies or lumens per watt so as to reduce the quantity of infrared heat produced per lumen of light.

- *Surface temperature:* Reduce temperature gradients between surfaces, objects, and occupants. Panel cooling and air circulation systems can cool the metal surfaces of offending systems.
- *Air temperature:* Reduce the room temperature to compensate for the excess radiant heat.
- *Selective control of energy:* Divert infrared energy away from the occupied zone by selective absorption, reflection, or transmission. Dichroic filters are available which allow visible radiation to be transmitted while reflecting infrared radiation. The reflected energy must then be controlled in some way so as not to allow it to eventually enter the occupied space.

REFERENCES

American Society of Heating, Refrigerating and Air-Conditioning Engineers, Inc. (ASHRAE). 1987. Handbook: *HVAC systems and applications;* 1988. Handbook: *Equipment;* 1989 Handbook: *Fundamentals;* Standard 90.1-1989. *Energy efficient design of new buildings except new low-rise residential buildings.* Atlanta.
Olivieri, Joseph B. 1987. *How to design heating-cooling comfort systems;* Fourth Edition. Troy, Mich. Business News Publishing Co.

PART THREE

COORDINATED SYSTEM DEVELOPMENT

Building design continues to be a challenge for improvement and excellence. At present, there is concern over the problem of consolidating and coordinating the significant technological achievements of the past century in order to incorporate them into our vocabulary of design and form. Much of the substance of the currently identified design vocabulary can be traced to the ideas and example of the Bauhaus (Germany) in the 1920s. This group taught designers to recognize the machine as a rising influence in western culture and building. Mass production, component assembly, functional simplicity, and modular attitude were shown to have logical and useful design implications.

During the several decades since such theories became sufficiently clarified to be recognized and accepted in our society, increasingly sophisticated technical developments have been introduced and perfected. Questions are raised concerning the significance of these and the changes they may portend for our approach to building design.

Some of these questions evolve from the fact that the architect today has greater freedom than ever before in reference to manipulation and control of sensory perception and comfort. But this freedom cannot be accepted as license for undisciplined design. The search for a sense of direction must begin by eliminating the question, How do you light and air-condition a preconceived building design?—replacing it with the question, How do you design a building that is in sympathy with the contemporary capability to manipulate and control the interior sensory environment? The building form that results from such a question should relate to both natural and artificially generated environmental influences. Part 3 will attempt to summarize and define some of the influencing factors.

CHAPTER 8

The Building as a Comprehensive Environmental System

Throughout history, two of the limiting criteria in building design have been to bring light and air into an enclosed space and to protect the interior from adverse external influences. Each society and culture has produced its own solutions to this problem, reflecting climatic demands as well as the technology and ingenuity of the time. The contemporary architect is also concerned with the problem of environmental performance. Continuing the long sequence of experiments and developments that have evolved throughout the history of building, the architect can, for example, develop his enclosure to utilize natural light, natural ventilation, and beneficial solar influences during colder periods of the year.

Several factors set the modern period apart from the past, however. This century has seen a continuing evolution of products and mechanical techniques for environmental control—electric lighting, mechanical air distribution systems, heating and cooling devices, and sound control. Viewed as parts of a comprehensive system, these techniques combine to provide a revolutionary new freedom for regulating the interior sensory environment. With this combination of tools the architect of the near-past found that the design of the exterior building shell was no longer *necessarily* subject to the uncompromising limitations of natural ventilation, daylighting, and solar control; and the occupants of the building found that their comfort and activities were increasingly independent of natural outdoor conditions related to climate, weather, or time of day.

Architecture reflects, in part, man's continuing attempt to establish a protected environment that approximates the conditions in which he is most comfortable and at ease, in spite of the fact that such comfort conditions appear only intermittently in nature. But rather than being approached as a simple correction of climatic deficiencies, the environmental control function of the building must be oriented toward the more extensive sensory demands of various occupant activities and experiences. At the same time, energy limitations and earth environmental aspects demand very careful use of resources in creating the built environment.

The building occupant perceives light as surface brightness and color; he absorbs heat from warmer surfaces and warmer air; and he himself emits heat to cooler surfaces and cooler air. He responds physiologically to humidity, to air motion, to radiation, and air freshness. He also responds to sound. A major function of the building, then, is to provide for all of these sensory responses concurrently—to establish and maintain order and harmony in the sensory environment.

THE DIAGRAMMATIC PLAN AS AN ORGANIZATIONAL TOOL

A *diagrammatic plan* is developed once the basic building program has been established. It is a useful planning tool for identifying, defining, and communicating among consultants the multiple and interdisciplinary demands involved in analysis of the man-made environment. This design step precedes the development of *schematic plans*, which begin to take on a definite— though preliminary—form and dimension. The function of the diagrammatic plan is to indicate basic relationships within and between various activities without implying either form or dimension. It is used to generate a definition of the building in terms of general performance requirements. In part, the diagrammatic plan serves as a relationship diagram that explores (without implying form) layout alternatives to be considered in subsequent planning stages, and it begins to identify the relative location of potential barriers and transitions.

Figure 8-1 indicates a representative diagrammatic plan for a grouping of five activities (identified simply as numbers for the purposes of this discussion). This type of graphic representation will facilitate a preliminary performance analysis of each separate activity as well as evaluation of barrier or transition relationships associated with each of these activities (between 1 and 2, 1 and 3, 2 and 3, and so on). Figure 8-2 takes this a step further and begins to explore the nature of the service distribution network.

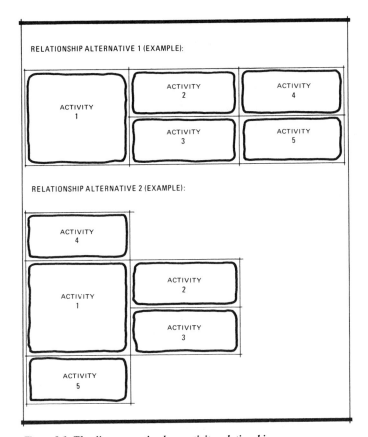

Figure 8-1. The diagrammatic plan: activity relationships.

The Comprehensive Performance Specification

Within the context of this book, the diagrammatic plan is a particularly useful organization tool for defining preliminary performance specifications to guide the ensuing system development. Continuing the example illustrated in figure 8-1, the pattern diagrams being studied here provide a framework for defining and evaluating the factors noted in tables 8-1 through 8-5.

SYSTEMS AND SUBSYSTEMS

As defined by the performance specification, the building is seen as a synthesis of several interacting systems and subsystems. There are four related but independently definable subsystem categories that must be successfully assimilated into the total environmental system design. These

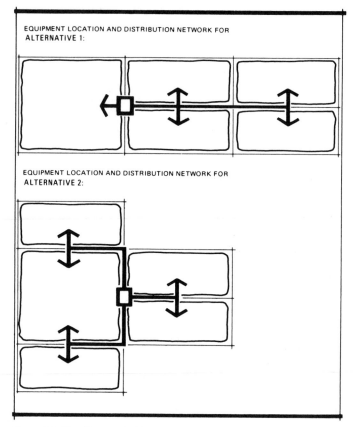

Figure 8-2. The diagrammatic plan: service distribution.

are the site systems that provide an environmental context or setting for the building, the subsystems that comprise the basic building enclosure, the subsystems that satisfy environmental and service demands within the building, and the subsystems that facilitate the distribution of energy and services to and throughout the building.

While most buildings are bid on prescriptive specifications rather than performance specifications, the planning stages should include the development of performance-related criteria.

The Site

In response to the performance requirements defined in table 8-1, site development can plan for environmental comfort by appropriate orientation and/or by incorporating natural screens and barriers for wind control, sun control, light control, and sound control.

Table 8-1. Organizational step 1: Develop environmental relationships with users.

Physical Environment	Social Environment	Task Environment
Orientation: Are users aware of form and artifacts?	Conversation: Are users aware of groups and individuals?	Comprehension: Are users aware of task detail and degree of visibility? Is task concentration enhanced by the environment?
Form analysis: Do users perceive form, objects, or space while attaching some meaning or feel to it (pleasant–unpleasant, spacious–confining, public–private, relaxed–tense)?	Occupant analysis: Do users remember a space because of the presence of other people—or because of the lack of other people? Are transactions and interacting promoted and/or enhanced by the space?	Task analysis: Do users perceive tasks in a way that facilitates fast, accurate decision making?
Functionalism: Is the meaning that users attach to a space relevant to the intended nature of the form, space, or activity? Does the space express its intended use?	Functionalism: Are the sensory contacts with people appropriate, desirable, or necessary, and are these contacts reinforced by the architectural organization?	Functionalism: Are the task functions reinforced by the building envelope and the activity area adjacencies?

Table 8-2. Organizational step 2: Develop the general building system (concept development).

Architectural-Environmental Parameters	Considerations
Building form, orientation, and location	Re: Daylight use Re: Solar gains Re: Prevailing winds Re: Precipitation Re: Views from surrounding areas Re: Utilities access Re: Users' access
Building materials and construction	Re: Response to daylight Re: Response to thermal conditions Re: Response to wind effects and infiltration Re: Response to noise abatement requirements Re: Response to integration of environmental systems Re: Response to electric lighting Re: Response to local, state, federal codes Re: Response to aesthetics Re: Response to intended image Re: Response to initial and operating costs Re: Response to human scale and use

Table 8-3. Organizational step 3: Define the nature of the activity space (concept development).

Space Concepts	Considerations (refer to discussions in parts 1 and 2)
Define basic opportunities and limitations regarding room volume and form	Re: Development of daylighting and natural ventilation effects Re: Development of appropriate natural amplification of sound

Table 8-3 (continued).

Space Concepts	Considerations (refer to discussions in parts 1 and 2)
	Re: Development of appropriate electronic amplification of sound
	Re: Development of appropriate reverberation control
Define basic opportunities and limitations regarding light source selection, reflector-refractor action, brightness and comfort control, housing and protective enclosures, positioning of luminaire elements	Re: Development of appropriate vector
	Re: Influences, spatial order, color whiteness
	Re: Development of appropriate task brightness and contrast
	Re: Development of appropriate visual comfort
Define basic opportunities and limitations regarding surface forms, materials, finishes, and assemblies	Re: Response to sonic reflection or absorption requirements
	Re: Development of appropriate interreflection of light
	Re: Development of appropriate mean radiant temperature
Define basic opportunities and limitations regarding air handling elements	Re: Response to environmental temperature and humidity control requirements
	Re: Response to air motion and distribution requirements
	Re: Air sanitation and odor control
Define basic opportunities and limitations regarding noise-generating devices	Re: Response to limitations in background noise
	Re: Development of appropriate noise screens to reinforce a sense of privacy
Define the essential nature of the control systems	Re: Lighting switches, dimmers, etc.
	Re: Electric service and convenience outlets
	Re: Microphone jacks, speaker jacks, etc.
	Re: Thermostats, etc.

Table 8-4. Organizational step 4: Define the nature of interior transitions and interior barriers.

Space Planning	*Considerations (refer to discussions in parts 1 and 2)*
Locate and define situations where continuity and/or circulation flow is required between adjacent spaces	Re: Development of appropriate continuity or successive contrast in regard to spatial lighting (brightness, color whiteness) Re: Development of appropriate continuity or successive contrast in regard to the sonic background Re: Development of appropriate continuity or successive contrast in regard to the thermal and atmospheric background temperature, humidity, MRT, air motion)
Locate barriers and define their essential nature	Re: Development of appropriate visual privacy Re: Development of appropriate sonic privacy and noise abatement Re: Development of appropriate thermal and/or humidity isolation Re: Development of fire barriers
Define basic opportunities and limitations regarding cosmetic requirements and functions associated with the barrier	Re: Ceiling-floor organization Re: Wall organization

Landscape and Surface Forms

Site vegetation and land forms can influence the immediate thermal environment of the building. These influences generally involve the diversion of storm winds, the channeling of cooling summer breezes, and sun shading. When localized exterior cooling is desired, fountains will provide some evaporative cooling of the air that passes through. For this reason, such water systems should be placed so that moderate-velocity prevailing summer breezes will be induced to pass through the fountain to cool the air prior to its passing into the outdoor living area.

Table 8-5. Organizational step 5: Define the nature of the environment.

Environmental Factors	Considerations (refer to discussions in parts 1 and 2)
Define basic opportunities and limitations regarding the heat source, fuel selection, and related apparatus requirements	Re: Response to heating and humidification requirements
Define basic opportunities and limitations regarding the cooling apparatus requirements	Re: Response to cooling and dehumidification requirements
Define basic opportunities and limitations regarding equipment space requirements and housing characteristics	Re: Location of heating apparatus (including chimney, fuel storage, etc.) Re: Location of cooling apparatus (including cooling tower, etc.) Re: Location of air-handling or fluid control apparatus Re: Special characteristics and code limitations
Define basic opportunities and limitations regarding the ventilation system	Re: Location of intakes and exhaust Re: Development of appropriate routing Re: Development of appropriate air sanitation and composition control (including odor control)
Define basic opportunities and limitations regarding the thermal distribution medium and network	Re: Development of appropriate routing Re: Development of appropriate zoning control

Similarly, wells or a body of water can be useful as a heat sink, replacing or supplementing the cooling tower as a means to reject condenser heat in the refrigeration cycle. Fountains and ponds can reinforce this action by evaporative cooling of the heat sink. At the same time, such retention ponds can serve as an additional fire water source. Landscape elements on a site can be used for thermal control assistance. Optimum position of these elements varies with prevailing wind patterns.

In a technical sense, these site factors influence the daytime luminous environment in a somewhat secondary way. Trees and screens that effectively control the thermal influence of solar radiation will also moderate the influences of direct sunlight and sky glare.

When daylight is to be utilized for single-story buildings, this is assisted by providing high-reflectance paved ground surfaces immediately adjacent to the building. However, this same reflector will also reflect significant solar heat into the interior space. For this reason, such reflector surfaces should be used with care, particularly on the west and south exposures.

Noise Control

When external noise cannot be muffled at the source, landscape barriers can provide some control within the site. These barriers generally involve either shielding or absorption (or both). The combination of trees, low foliage, and ground cover provide noise attenuation when significant masses of such absorbing vegetation are involved. Generally, a 500–1,000-foot depth of such foliage is required to diminish the intensity of normal traffic noise adequately. While relatively thin barriers serve as an effective visual barrier or sun screen, a sonic barrier must be of much more significant dimensions.

Table 8-6. Desirable site utilization for north-temperate climatic conditions.

---

1. To facilitate maximum exposure to the sun during prolonged winter periods, utilize warm slopes for building sites in colder regions.

2. Where it is desirable to provide natural summer cooling, utilize the lower portion of windward slopes. Furthermore, in order to induce penetration of prevailing summer breezes, openings should be placed to admit ventilation air on the windward side of the building, with exhaust outlets placed on the leeward side.

As a related factor, in order to facilitate natural internal cooling action during warm periods, minimize blockage of prevailing summer breezes. (Usually this means that dense site screening should be curtailed on the south and southwest exposures.)

3. Wind screening is desirable on the windward side of the building during cold periods. (Usually this consideration applies to the north and northwest exposures.)

4. Utilize evergreens for wind-screening purposes. Utilize deciduous trees for sun-shading purposes.

5. If possible, locate building so the available fully developed shade trees will provide shading on the east and west sides of low buildings. Similar considerations apply for the location of outdoor living areas.

6. Paving should be minimized immediately adjacent to the building. Where possible, vegetation should be used in this location to absorb (rather than reflect) solar energy. The critical west and southwest exposures are most likely to produce significant reflected energy during periods of peak solar heat gain.

7. Walks should be shielded from winter winds and summer sun.

Table 8-6 describes a variety of conditions which may be used to maximize site use in north-temperate climates.

More effective control may be provided with shielding (fig. 8-3). In order for such a shield to be effective it must have significant mass and it must be impervious to air flow. The shield must also be of sufficient height to screen the line of sight between the noise source and the receiver. This technique casts a *sonic shadow* over areas bordering the noise source. This deflection of sound creates an area of reduced noise intensity (up to 25dB) within the *shadow zone*.

Site Service Networks

The building environmental and service systems depend on the provision of adequate site systems for electric power, water supply, gas, communications, and liquid disposal. These utilities and services can be distributed in walk-through tunnels that are fully accessible for maintenance. However,

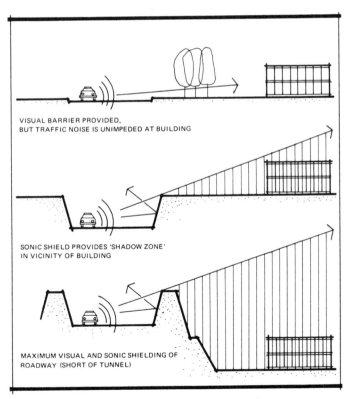

Figure 8-3. Development of topography for noise control.

this method of distribution tends to be costly and is generally limited to compact projects.

More generally useful are concrete or tile trenches and metal conduit enclosures, both placed under the sidewalk (fig. 8-4). The paving can then be designed to be lifted so the piping is accessible as required for maintenance and future connections. This access should be available without disturbing the landscape.

The Enclosure System

Immediately related to the aspect of site development is the development of the enclosure system that defines the external character and form of the

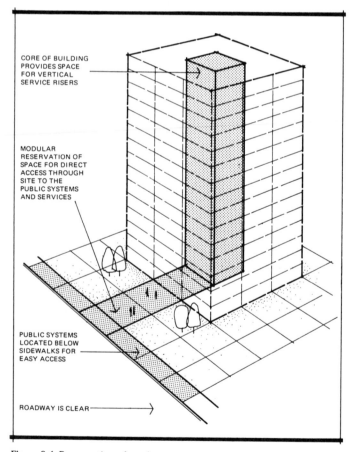

CORE OF BUILDING PROVIDES SPACE FOR VERTICAL SERVICE RISERS

MODULAR RESERVATION OF SPACE FOR DIRECT ACCESS THROUGH SITE TO THE PUBLIC SYSTEMS AND SERVICES

PUBLIC SYSTEMS LOCATED BELOW SIDEWALKS FOR EASY ACCESS

ROADWAY IS CLEAR

Figure 8-4. Preservation of service access.

building. In one sense, the enclosure serves as a barrier to protect the interior space from hostile or inappropriate external influences. At the same time, the external environment contains many favorable influences; so the enclosure may serve more precisely as a selective filter that allows these favorable influences to penetrate. The appropriateness of the enclosure system depends on its success in intercepting, modifying, or otherwise regulating the penetration of external influences into each activity-oriented interior space.

Consideration should be given to the fact that extensive use of glass enclosure systems will require tinted or light-reflecting techniques to make both solar loads and interior brightness control discernible. The result may be a feeling from the interior spaces that exterior conditions are constantly overcast. Appropriate interior electric lighting techniques can be used to minimize an overriding impression of gloom. Architectural shading systems or use of smaller areas of clearer glass as opposed to heavily tinted glass can enable a more realistic view outdoors. Table 8-7 lists a variety of subsystems and corresponding performance criteria to help ensure a successful building enclosure system.

Decisions Regarding Dominance of Natural or Mechanical Influences

As interior activities are studied, it is possible to make decisions regarding the appropriateness of various influences in the natural (exterior) environment. At this point, these influences can be either utilized or excluded through the methods discussed in chapters 5 and 7. Tables 8-8 and 8-9 deal with the general situations and conditions in which natural and electrical-mechanical alternatives become relevant.

Table 8-7. The enclosure system.

Subsystems	*Environmental Performance Criteria*
The roofing subsystem The exterior wall subsystem The foundation-floor subsystem	1. Control solar heat penetration through openings 2. Control transmission of light through openings 3. Control heat conduction through the assembly 4. Control condensation within the assembly 5. Control the mean radiant temperature effect of the interior faces of the enclosure 6. Control the transmission of external noise 7. Control the penetration and exhaust of ventilation air

Table 8-8. Impact of lighting conditions on environment.

Daylight-View Implications	*Electric Light Implications*
Narrow, rectilinear building shapes	Spatial enhancement (use electric lighting in concert with architecture)
Higher ceilings	
Extensive glazing areas (traditional windows along with skylights and roof monitors)	Psychological enhancement (use electric lighting to influence occupants in a positive, subjective manner—pleasantness, spacious-ness, visual clarity, etc.)
Diffuse, highly reflective interior surfaces	
Direct solar shading devices (archi-tectural overhangs, shades, baffles, etc.)	Task enhancement (use electric lighting to promote optimum visual performance of occupants)
Atria (to introduce sense of daylight to interior areas; to provide "exterior" view for interior areas)	Brightness balancing of daylighting
	Control integration with daylight
	Control to provide multi-use of space
Brightness balancing (use electric light to provide comfortable transition from perimeter daylighted areas to interior areas)	Focal center development
	Integration with other disciplines (mechanical, structural, architec-tural, interior design, signage)
Glare and washout potential of elec-tronic office tasks (may require use of low transmittance glazing and/or shades and/or special treatment of video display terminal screens)	Maintenance (owning and operating conditions necessary for optimum use)
	Energy Consumption
Energy/Consumption (thermal conditions)	
Acceptance of varied lighting conditions	

Thermal Performance of Building Forms

When developing the enclosure form itself, it is significant that each face of the building is subject to somewhat unique climatic influences related to wind, sun, and conductive interchanges with the air mass. These influences are discussed in chapter 7, and the more significant solar patterns are briefly summarized in figure 8-5.

Performance of Compact and Finger-Type Building Plans

Beyond the implications of differing exposures associated with individual building faces, a more general relationship exists between the essential form of the building and the operating performance of balancing mechanical devices. For example, a *compact* building form is one that encloses a

Table 8-9. Impact of thermal conditions on environment.

Natural Ventilation Implications	*Mechanical Ventilation Implications*
Narrow, rectilinear building shapes	Complicated architectural shapes can be accommodated
Operable sashes	Constant thermal conditions (temperature, humidity, and air motion control can be limited to narrow ranges)
Suitable siting for cross-ventilation	
Location with suitable climate (clean air, reasonable temperature and humidity)	Controlled sonic environment (little or no external noise sources)
Acceptance of varied thermal conditions (day-to-day, hour-to-hour variations)	Adjustable (to accommodate increasing or decreasing demands due to use of VDT's and due to occupancy loading)
Reduced sonic control (distracting external sounds or wind noise)	Integration with other disciplines (lighting, structural, architectural, interior design)
Daylight availability	Maintenance
	Energy consumption (redistribution of heat in high-heat areas to low-heat areas)

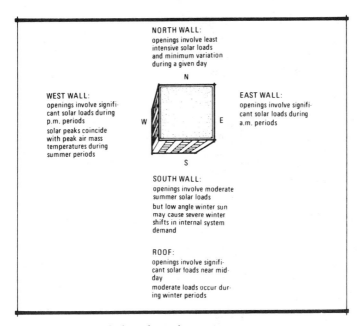

NORTH WALL:
openings involve least intensive solar loads and minimum variation during a given day

WEST WALL:
openings involve significant solar loads during p.m. periods
solar peaks coincide with peak air mass temperatures during summer periods

EAST WALL:
openings involve significant solar loads during a.m. periods

SOUTH WALL:
openings involve moderate summer solar loads
but low angle winter sun may cause severe winter shifts in internal system demand

ROOF:
openings involve significant solar loads near mid-day
moderate loads occur during winter periods

Figure 8-5. Solar loads through openings.

maximum floor area with minimum perimeter footage (a square is the most compact rectangular form). This minimizing of the interior-exterior contact surface will generally improve the thermodynamic characteristics of the enclosure in both cold and hot-arid situations. However, compact building forms tend to be somewhat limited in their ability to respond to moderate and humid climatic conditions, when ventilation flow is decisive in maintaining occupant comfort. As a result of this limitation, these building forms will present more extensive cooling demands during moderate and humid periods.

Irregular *finger-type* or decentralized building plans require a greater perimeter footage to enclose the same floor area. These forms will tend to permit a better natural interaction during moderate and humid periods because ventilation air can generally be led to circulate naturally through the narrower building sections (by cross-flow). However, these finger-type forms offer a poorer response during cold periods because of the greater interior-exterior conduction surface. As a result, these building forms will present more extensive heating demands during winter periods.

Since north temperature regions tend to evolve a variety of external conditions during the various seasons of the year, both of these basic building form alternatives will be somewhat limited in ability to respond naturally to thermal influences during some portions of the year. For this reason, a balancing mechanical system must become an integral part of the concept; and the relative size and significance of individual mechanical components will vary with the building form.

Comparative Performance of Building Shapes

Figure 8-6 summarizes heat gain and heat loss performances for several typical building configurations. Notice that as the building becomes larger (vertical or horizontal expansion), the heating and cooling systems will operate more efficiently because the relative influence of heat transfers through roof or wall surfaces is reduced.

When the building shell involves minimum window openings, conductive heat transfer is a dominant thermal influence. Therefore both heating and cooling demands are minimized with compact building shapes. On the other hand, when significant glass areas are involved, structures that are elongated in an east-west direction may tend to minimize cooling demand. This is particularly true when internal heat sources are negligible. This conclusion tends to be complicated somewhat when a major internal heat load is present (such as a high level electric lighting load). Since orientation along the east-west axis still leaves the south face exposed to significant direct radiation on clear winter days when the solar altitude is low, winter solar heat (in addition to significant lighting heat) can produce overheated interior conditions, even on a very cold day. For this reason, glass in the south wall of commercial or institutional buildings should be used with restraint, unless adequate shading is provided.

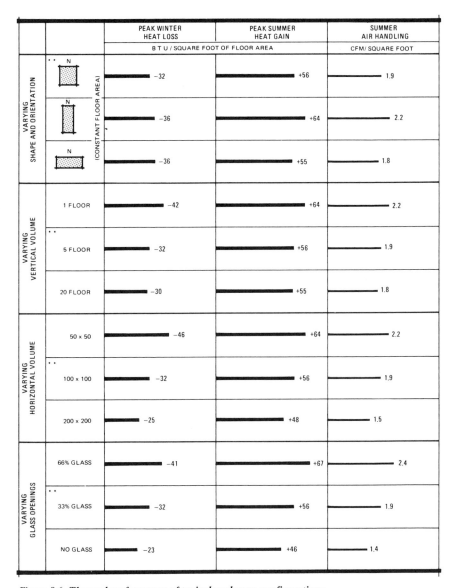

		PEAK WINTER HEAT LOSS	PEAK SUMMER HEAT GAIN	SUMMER AIR HANDLING
		B T U / SQUARE FOOT OF FLOOR AREA		CFM / SQUARE FOOT
VARYING SHAPE AND ORIENTATION (CONSTANT FLOOR AREA)	N	−32	+56	1.9
	N	−36	+64	2.2
	N	−36	+55	1.8
VARYING VERTICAL VOLUME	1 FLOOR	−42	+64	2.2
	5 FLOOR	−32	+56	1.9
	20 FLOOR	−30	+55	1.8
VARYING HORIZONTAL VOLUME	50 × 50	−46	+64	2.2
	100 × 100	−32	+56	1.9
	200 × 200	−25	+48	1.5
VARYING GLASS OPENINGS	66% GLASS	−41	+67	2.4
	33% GLASS	−32	+56	1.9
	NO GLASS	−23	+46	1.4

Figure 8-6. Thermal performance of typical enclosure configurations.

Related Internal Planning Relationships

The previous comments have referred primarily to the nature of exterior walls as filters for regulating the penetration of external environmental influences. However, as a related technique, a selective placement of activities and rooms can become useful as a reinforcing device in both sonic and thermal development.

Insulation from Noise Influences

The enclosure shell should insulate the interior space from external noises associated with traffic, aircraft, rail activity, stationary machinery, playground activity, and so on. The shell should be sealed to prevent airborne noise penetration, and it should provide suitable mass or structural discontinuity to inhibit the penetration of external noises (see chapter 6).

When the enclosure itself may be inadequate to provide a completely effective barrier, critical interior areas should be further shielded from troublesome external noise sources. As an example, the bedrooms of a home can be shielded from external road noises and playground noises by using other, less critical interior spaces as separating *buffers* (fig. 8-7).

In commercial space, the division and segregation of space for acoustical reasons is also important. Major circulation axes may act as buffers between more industrial-type spaces (test labs, model shop, research labs, and so

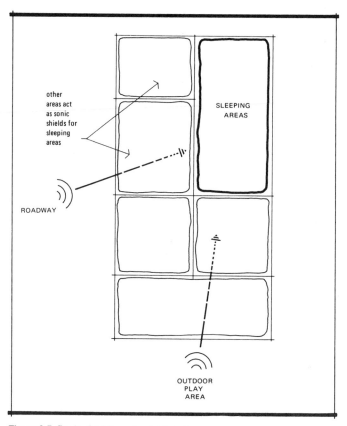

Figure 8-7. Sonic shielding of quiet interior areas.

forth) and office functions (such as purchasing, corporate communication, and the cafeteria).

Insulation or Exposure of Thermal Influences

Additional layout considerations evolve from thermal influences. General recommendations are summarized in table 8-10 and among the final items in table 8-11. This latter table also provides a more general summary of thermal considerations associated with the enclosure.

The Service Distribution Network

The initial function of the mechanical service system is to compensate for environmental deficiencies or excesses caused by inconsistent interaction between the building shell and the surrounding natural environment. In the temperate regions of North America, the required service distribution networks and related process equipment are often relatively extensive and they tend to make rather extensive demands on the physical form of the building.

As a result, these service networks are somewhat expressive of the nature of the building. They define the flow of energy and services; and in this sense the network should be a disciplined and logical expression that facilitates both accessibility and flexibility for change.

Essentially the concept of the building will reflect a number of system decisions that can be defined within the context of the diagrammatic plan (see figure 8-2 and table 8-5). How are the services distributed vertically? How are they distributed horizontally? Is there to be a *centralized* or *decentralized* distribution network? And finally, what space and linkages are required for the major process equipment?

Table 8-10. Recommended solar orientation for various rooms.

	N	NE	E	SE	S	SW	W	NW
Work space with exterior exposure	X	X						X
Recreational and lounge areas with exterior exposure				X	X	X	(X)	
Courtyards and terraces			(X)	X	X	X	(X)	

NOTE: The American Public Health Association committee on the hygiene of housing has recommended: "at the winter solstice, at least one-half of the habitable rooms of a dwelling should have a penetration of direct sunlight of one-half hour's duration during the noon hours when the sun is at its maximum intensity." (This recommendation refers primarily to the germicidal action of solar radiation.)

Table 8-11. Desirable characteristics for exposed enclosures in north temperate climate.

1. In terms of the utilization of mechanical heating and cooling capacity, *compact* building forms (those with a minimum *shell area to enclosed area* ratio) will tend to be more optimum energy-conserving building shapes when the enclosure is exposed to extreme seasonal conditions (cold or hot-arid). However, significant mechanical ventilation and cooling of interior core area will be required during moderate and humid periods because *compact* building forms tend to be limited in their ability to interact naturally with the external environment.

2. In contrast with the previous comment relating to *compact* building forms, penetration of natural light, natural cross-ventilation, and free air flow through interior spaces all imply the development of narrow building elements and *open plan* interior partitioning.

3. When only minimum to moderate internal heat source concentrations are present in the interior, building shapes that are elongated along the east-west axis are generally preferable in that they tend to minimize solar heat gain during peak summer periods.

4. Utilize light to medium exterior finishes for exterior walls and roof areas that are shielded from the summer sun.

5. When possible, shading devices should be external—and they should be detached from the structure in the sense that they should be exposed to maximum cooling by wind convection.

6. Wall and roof construction should have good insulation characteristics, be non-porous, and be resistant to freeze-thaw action. Construction should resist moisture penetration. (These characteristics tend to be more critical on the windward north and west exposures.)

7. Where appropriate, roof construction in snow areas should utilize simple sloped roof forms to facilitate snow removal by wind action. Simple roof forms will minimize snow accumulation and ice-filled gutters.

8. When buildings are placed on open sites, plan development should generally open to the south or southeast and should be relatively closed to the north and west (to screen storm winds and late afternoon sun). Therefore, window areas on the north and west exposures should be somewhat limited.

9. When buildings are placed on open sites, noninhabited spaces should generally be placed in the west and southwest portions of the plan to serve as insulators against solar effects.

10. Courtyards and outdoor lounge areas should be oriented toward the south in order to maximize the use of these areas as an extension of interior spaces during periods of moderate to cool weather. Shading of the west sun is desirable during summer periods.

11. Interior heat-producing and humidity-producing areas should be separated from other occupied areas.

Mechanical Service Space

Most buildings are composed of three types of interior space. One type, the useful or functional space that is available for the essential human activities that take place within the building, includes offices, classrooms, lounges, conference rooms, eating space, and so on. These spaces provide the predominant framework for placement of terminal air diffusers, luminaires, acoustical absorption, and other mechanical service elements (to be discussed in the succeeding section).

The second type of space is the major human circulation and exit routing that provides physical access to and from human activity areas. This will include entrance lobbies, stairs, elevators, and corridors, and usually the personal or building services that are immediately accessible from the circulation route (areas such as public toilet rooms and janitorial spaces). This human circulation path provides a potential framework for the organization and placement of centralized distribution networks to and from the functional spaces. Duct and other service routes can parallel horizontal corridors and vertical stair towers.

A third type of space is *mechanical service* space. This will include some areas that require intermittent human access (such as equipment rooms, fan rooms, and so on) and other areas that are somewhat remote from immediate human access (such as vertical and horizontal trunk duct space). As a category, however, the placement and sizing of mechanical service spaces are generally dependent on equipment needs rather than human needs.

Each of these three space categories is essentially separate and definable within the building design.

Layout and Expansion of Mechanical Service Spaces

Table 8-12 summarizes some basic space allowances that are useful for estimating purposes during the initial schematic design phase. When significant future expansion is anticipated, equipment and plant design should be based on carefully defined increments of growth. These increments (equipment modules) can then be installed in stages as required, utilizing previously reserved areas and a common distribution network.

The initial stage of a central mechanical service plant, for example, may include two equipment modules—one to provide basic capacity and one to provide 100 percent standby capacity. As the building size is doubled, a third equipment module is added. In this way, a 100 percent growth in load requires only a 50 percent expansion of service capacity (because one equipment module is always available as standby to take over for one of the other two). As this process continues in future expansions, an efficient system of service units evolves that is capable of responding at various capacity levels to meet varying load conditions.

Table 8-12. Mechanical space allowances for use in preliminary planning.

Centralized mechanical equipment rooms	Allow approximately 5-8% of the gross floor area for refrigeration equipment, heating devices, and related pumps.
	When positioning boiler and chiller devices, allow adequate space immediately adjacent to the equipment to permit removal and maintenance of the tubes.
	Generally this space allowance approximates the length of the device.
	Allow equipment room ceiling heights of approximately 13-18 ft.
	Allow approximately 2% of the gross floor area for air handling components.
	Consider thermal and acoustical insulation requirements when developing wall, ceiling, and floor details for all equipment rooms.
Cooling towers	Allow approximately 1 sq. ft. of roof area for each 400 sq. ft. of gross building area.
	Height allowances range from approximately 13 ft. (for small buildings) to approximately 40 ft. (for very large buildings or building groups).
	Structural load allowances range from 120–200 psf with water.
	Allow approximately 4 ft. of access space below the unit for access and piping.
	Consider acoustical isolation requirements when developing adjacent roof and wall details; also consider the need for resilient cushioning to prevent structural transmission of vibrations.
Chimneys	For gravity systems, place the natural draft chimney 100 ft. or more from fan room air intakes and cooling tower intakes.
	Forced-draft chimneys can generally be placed in closer proximity, provided exhaust air flow is directed away from air intakes.

Split-package mechanical equipment	Typical 5–20 ton limit per unit. Locate air-cooled condenser units near the perimeter because of difficulties in overcoming duct friction. Remote interior evaporator is linked to the condenser with piping: • Maximum horizontal distance to interior evaporator: approximately 60 ft. • Maximum vertical distance to interior evaporator: approximately 30 ft. (Therefore, when multiple units are to be used, consider (1) 5-floor vertical zones, with the mechanical unit located at the center floor; or (2) utilization of one unit to serve each moderate-sized floor).
Roof-top, multizone packages	Typical 5–60 ton limit per unit. These are self-contained heating, cooling, and air-handling devices (excluding ducts). Height allowances range from approximately 5 ft. (for smaller units) to 10 ft. (for larger units). Roof area allowances range approximately 4 sq. ft. per ton of capacity (for larger, more efficient packages) to 6 sq. ft. per ton (for smaller packages).
Centralized mechanical service zones	One centralized mechanical equipment room will service approximately 8–20 floors of a high rise building. The more floors served, the larger the duct shafts and the greater the mechanical equipment room volume at each single location.

(continued)

Table 8-12 (continued).

Air duct allowances	Horizontal duct allowances: (for estimating *drop ceiling* cross-sections for corridors, etc.)
	• Allow approximately 1–2 sq. ft. of cross-section for supply ducts for each 1,000 sq. ft. of occupied space. (This will vary, depending on the velocity of air in the trunk; i.e., higher velocity air flow requires smaller ducts. The allowance will also vary with the complexity of the duct layout and with the air change rate required; i.e., simplified duct layouts and reduced heat gains will reduce duct size requirements.)
	• The supply duct allowance must be duplicated for the return duct system.
	• Consider the fact that return air velocities may be lower than the supply requiring larger duct allowances.
	Vertical shaft allowances:
	• Shaft allowances for a given floor and for a group of floors will be based on the same supply and return allowances considered above.
	Shaft and duct allowances may need to be increased 2 times or more to compensate for air flow inefficiencies such as excessive elbows, angles, boot fittings, duct take-offs, etc. For this reason, complex and irregular duct patterns should be avoided, and attention should be given to the greater efficiency of round ducts over square and rectangular, the least efficient.

The Mechanical Service Zone

Mechanical service zones are those limited portions of a building that can be defined as *detachable* or semi-independent mechanical operating units. These constitute one of the organizing elements in building planning. Generally, these zones are repeatable units, such as a floor of a building, a group of floors, or some other definable structural segment. In each case, the zone is served with electric power, communications, plumbing, and the HVAC services necessary to facilitate the intended activities within that zone.

For some of these services, the zone may be self-sustaining. For example, package-processing equipment may be located integrally within the zone. In other cases, all services are brought to the zone from some centralized remote equipment location. But in either case, each mechanical service zone involves a service trunk and entrance that serves as a point of access and control and a comprehensive internal distribution network that provides for internal service access to all parts of the zone.

The definition of this service zone (or zones) is an important aspect of planning because it establishes a discipline within which the various electrical and mechanical advisors can work in harmony with the architectural design. Without such a definition, this aspect of building design tends to remain fragmented and undisciplined.

Centralized Distribution Networks

The service distribution networks must physically relate to each of the three types of interior space (functional space, human circulation or access space, and mechanical equipment areas). This fact can become a useful discipline in network planning, because when circulation and access evolve around the concept of a core-corridor system, one logical distribution path is created for vertical (e.g., elevator, escalator) and horizontal (e.g., corridors) human circulation routes (see fig. 8-8).

According to this theory, the access core that provides a continuous vertical right-of-way for elevators and stairs will also provide a manageable vertical route for major liquid, air, and electric power distribution risers (fig. 8-9). Mechanical equipment and devices will be clustered in a convenient manner near this core (see fig. 8-10). Similarly, the major horizontal mains or trunks will parallel the major corridor systems (as in a drop ceiling). Feeders can then run off from these trunks as necessary to serve individual or clustered rooms in a manner not unlike the way that doorways provide access to these same rooms from the corridors. This procedure will ensure an organized and accessible right-of-way to both near and remote activity areas.

Decentralized Distribution Networks

An alternative routing of services can parallel the structural system. Because of the inherent decentralization of the structural columns, this

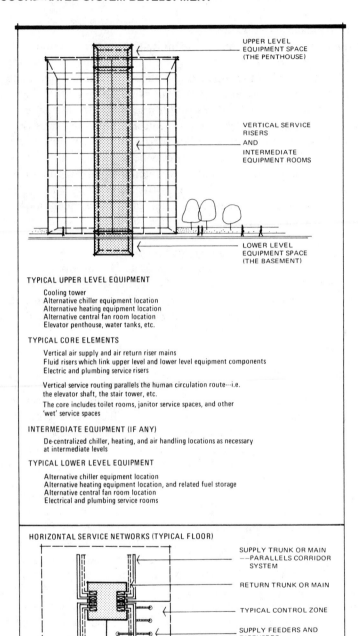

TYPICAL UPPER LEVEL EQUIPMENT

Cooling tower
Alternative chiller equipment location
Alternative heating equipment location
Alternative central fan room location
Elevator penthouse, water tanks, etc.

TYPICAL CORE ELEMENTS

Vertical air supply and air return riser mains
Fluid risers which link upper level and lower level equipment components
Electric and plumbing service risers

Vertical service routing parallels the human circulation route---i.e.
the elevator shaft, the stair tower, etc.

The core includes toilet rooms, janitor service spaces, and other
'wet' service spaces

INTERMEDIATE EQUIPMENT (IF ANY)

De-centralized chiller, heating, and air handling locations as necessary
at intermediate levels

TYPICAL LOWER LEVEL EQUIPMENT

Alternative chiller equipment location
Alternative heating equipment location, and related fuel storage
Alternative central fan room location
Electrical and plumbing service rooms

Figure 8-8. The centralized mechanical distribution network.

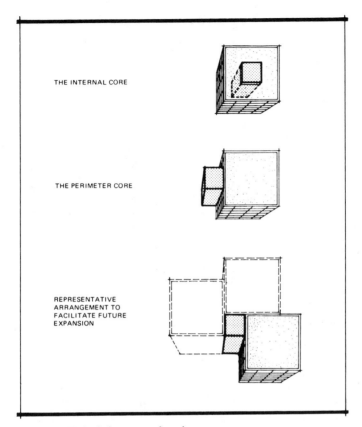

THE INTERNAL CORE

THE PERIMETER CORE

REPRESENTATIVE
ARRANGEMENT TO
FACILITATE FUTURE
EXPANSION

Figure 8-9. Typical placement of service cores.

concept will naturally produce a decentralized vertical distribution network.

According to this theory, vertical routing is contained within the column, while horizontal routing of mains and trunks is contained within the major structural beams. In such cases, the need for service space will probably lead to a somewhat more complex definition of the structure than is likely when only simple structural loading is involved. For example, accessible spaces may be defined by double beams and columns, or structural members may serve a dual capacity as a riser shaft and duct (fig. 8-11).

Comprehensive Interior Environmental Systems

The basic activity unit is generally a room or a similar spatial subdivision. The specific needs of this room will vary according to the nature of the occupancy or activity. But in order for the space unit to function successfully, it must perform a number of environmental and service functions that are summarized in table 8-13.

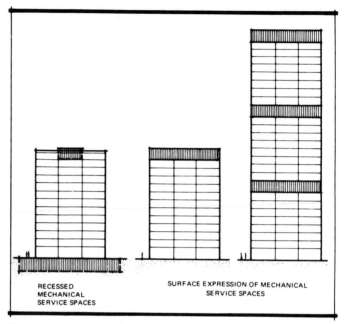

Figure 8-10. Typical massing of mechanical service spaces.

If the interior spatial unit is to function in an effective way, all aspects of the sensory environment (luminous, thermal, and sonic) must be satisfied *concurrently* as integral parts of a total environmental setting or background. While the individual disciplines of light, heat, and sound can be discussed as somewhat fragmented concepts, the actual interior spatial unit must represent a synthesis.

System Components and Assembly

Comprehensive interior systems may be developed for a specific building project. They may also reflect contemporary methods of industrial production and assembly, particularly for high production *vernacular* building types such as office buildings, schools, stores, and laboratories. Each of these is generally characterized by the repetitive nature of the interior spaces and by the requirement of flexibility to facilitate change and growth. The building requirements for such structures have stimulated recent efforts to provide construction and assembly solutions in four related areas:

1. The use of assembly line techniques to provide for subassembly of components at a remote location and rapid integration of effective systems on the job site

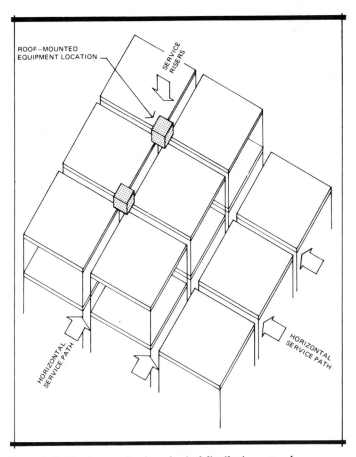

Figure 8-11. The decentralized mechanical distribution network.

2. The use of various modular approaches in an attempt to solve the distribution problems associated with industrialized components (reflecting the growing practice of manufacturing and stocking building components without knowledge of who will buy them or where they will be used)

3. The solution to the problems associated with joining separate components and subassemblies and the reduction of special skills or tools required for field assembly

4. The fundamental need to provide flexibility for spatial variation

The techniques of mass production offer the possibility (if not the immediate realization) of precision components and lower building costs if the building process can be organized to utilize the unique potential of the production line while recognizing its limitations. A successful building approach

Table 8-13. The interior system.

Subsystems	Performance Functions
The partition subsystem The ceiling-floor subsystem	1. Control air-borne sound transmission between spaces. 2. Prevent structural-borne sound transmission between spaces. 3. Provide appropriate sound control within the occupied space (i.e., absorption attenuation or reflection). 4. Provide appropriate intensity, color, and distribution of light in the occupied space (i.e., light sources, reflectors, lenses, brightness control devices). 5. Provide appropriate light reflection characteristics within the occupied space. 6. Provide air or water distribution for thermal processing within the space (i.e., ducts, piping, valves, dampers, mixing boxes, etc.). 7. Provide for electrical distribution and control (i.e., wireways, outlets, switches, etc.). 8. Provide for electrical distribution and control (if applicable for wet spaces). 9. Provide for distribution of communication and electronics services (i.e., wireways, outlets, etc.).

(for interior environmental systems) must be developed to utilize repetitive units produced on a standard assembly line. Alternatively, production line techniques need to be utilized to permit economical unit variations in form, dimension, texture, and color. In either case, industrialization requires a minimizing of the number of different components or assemblies required, of the physical steps required of both factory and field labor, and of the number of actual components to be handled in the field. Table 8-14 lists a variety of possible advantages of mass-produced interior system components.

Modular Development

With the development of sophisticated assembly line procedures has come considerable experimentation in the techniques of dimensional coordination and modular repetition. The *standard dimensional module* is an attempt to provide a horizontal and vertical dimensional framework (planning grid) through which the various building design disciplines can corre-

Table 8-14. Potential advantages of industrialized interior systems.

1. Utilization of pretested and coordinated components facilitates the development of predetermined quality levels and performance standards.

2. Utilization of dimensional and performance disciplines facilitates the ability to preselect compatible structural components, partition-enclosure components, and environmental control components—and this facilitates a reduction in field design and engineering time.

3. Predictability of erection procedures facilitates more accurately determined cost projections and bidding procedures.

4. Utilization of standardized components facilitates speed or delivery and improved construction scheduling.

5. Utilization of pretested and coordinated assemblies facilitates erection procedures, provides improved component consistency, and generally provides improved installation economies over conventional construction of comparable quality.

6. Emphasis on production compatibility and interchangeability facilitates flexibility for spatial rearrangement and modification as activity requirements evolve and change.

late components and assemblies. This approach utilizes components such as structural members of a consistent and common length and partition units and electrical-mechanical elements of related size. Figures 8-12 through 8-14 illustrate this basic approach to industrialized building design.

The *human-use module* is a spatial approach rather than one oriented toward component assembly. This module could be a completely prefabricated bathroom, kitchen, office, or classroom—possibly shipped complete to a job site for incorporation as a finished product. It includes all elements necessary to perform a given task or activity. (In theory, the automobile and the airplane are examples of mass-produced, regularly variable human-use modules, as is an office workstation.)

The Mechanically Coordinated Module

The term *mechanically coordinated module* refers to a standard spatial unit that includes lighting, air supply, air return, and appropriate sound control together with the necessary sanitary, communication, and power services (fig. 8-13).

Each individual environmental and service component is dimensionally related to the modular floor, wall, and ceiling systems; and the combination must be compatible with the structural and space furnishing systems. The resulting assembly may also perform a cosmetic function by correlating (and possibly concealing) the somewhat complex assembly of components and networks.

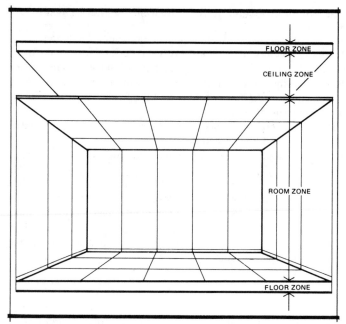

Figure 8-12. The system matrix grid.

The mechanically coordinated module, then, is the basic self-contained unit of repetitive interior assembly. It should be sufficiently flexible to provide maximum design freedom for the initial designer or the future tenant. And yet the systems should be sufficiently disciplined in a technical sense to ensure that the resulting rooms can be made to function in a logical and efficient way.

When this approach is effectively pursued, the mechanically coordinated module becomes a basic subsystem in interior planning. Any combination of modules can be isolated and identified as an independent room; and the resulting space will automatically include the capability to provide for luminous, thermal, sonic, and other service needs.

On-Site Assembly

An important aspect of interior environmental systems is the problem of on-site assembly. Methods of joining the various modules or components is a critical consideration—one with a potential for causing conflict and construction delays. A desirable objective for any industrialized building project is an extremely simple and well-defined method of joining, one that reduces

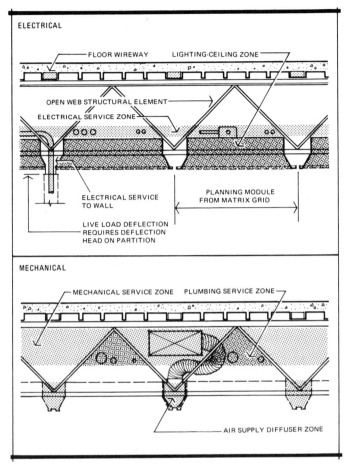

Figure 8-13. The planning matrix: space allocation in the ceiling zone (steel).

to a minimum the special skill or tools required for field assembly. Complex connections, excessive maneuvering, or field alterations quickly submerge the economic advantages of industrialization.

Component and Joint Tolerances

A basic problem in industrialization (and perhaps the one most difficult to solve) is that the size and shape of each joint detail must be standardized within close tolerances. There is some latitude in the stringency of this requirement; but this factor seems to have become progressively more

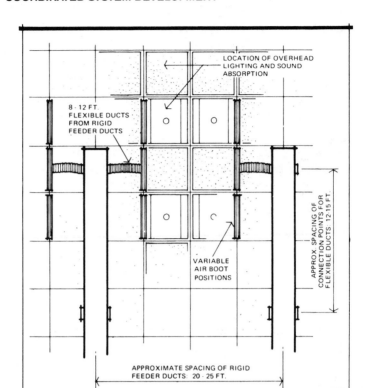

Figure 8-14. The planning matrix: typical air-handling development.

precise and limited with the advance from the somewhat flexible character of masonry joints to the close tolerances that are associated with many metal-to-metal joints, electrical contact joints, or fluid service joints.

When the joint is taken into account, the somewhat rigid modular dimension may not necessarily equal the outside dimension of the major elements or components. The modular dimension is determined by the assembled component-joint combination. Similar to the length of a wavecycle, the module dimension is measured from a certain point—for example, mid-joint—to the next repeated identical point—mid-joint again. Tolerances must therefore be established for components and joints both separately and together.

Labor Jurisdiction

A related consideration in the design and installation of comprehensive environmental systems concerns disputes that may arise regarding the work

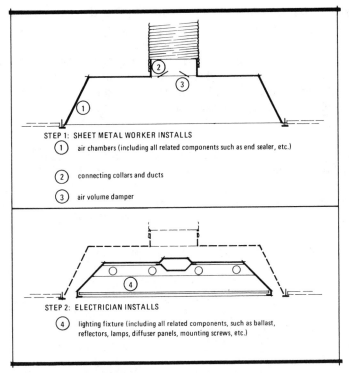

STEP 1: SHEET METAL WORKER INSTALLS
(1) air chambers (including all related components such as end sealer, etc.)

(2) connecting collars and ducts

(3) air volume damper

STEP 2: ELECTRICIAN INSTALLS
(4) lighting fixture (including all related components, such as ballast, reflectors, lamps, diffuser panels, mounting screws, etc.)

Figure 8-15. Typical on-site jurisdiction for installation of multi-function assemblies.

to be done by various trades. As an example of attempts to clarify this situation, the following agreements generally apply to disputes in the installation of ceiling elements. (Also see fig. 8-15.)

1. The installation of typical acoustical ceilings is the work of the carpenter trades. The electrical trades will install all related wiring devices.

2. When luminaires fit into a grid system, the grid systems that support both acoustical tile and luminaires are installed by the carpenter trades to the extent of providing parallel tees on two sides of the luminaires installed by the electrical trades.

The electricians will install all tees and hangers that support only luminaries. They will also mechanically secure and ground all fixtures.

3. When the luminaires are suspended from a grid ceiling, the electrical trades will install all supporting hangers, install the luminaires, and install all supporting members. The carpenter trades will install all acoustical materials.

4. For a luminous ceiling, the lamp channels on the upper ceiling are installed by the electrical trades. Beyond this, the general agreement is that if

the diffuser panels and supporting grid are suspended from the electric lamp channels, the suspension system is installed by the electrical trades. If the diffuser panels and supporting grid are suspended independently of the luminaires, the suspension system is installed by the carpenter trades.

The controlling trade will then install the diffuser panels themselves. However, if an acoustical border or other acoustical panels are involved, these panels and related suspension elements will be installed by the carpenter trades.

Combinations of services conventionally handled by differing jurisdictions of building trades can result in problems that add to project costs in certain municipalities. While a fluorescent luminaire incorporating an air supply system can be readily designed and manufactured, separation of the functions into components for separate trades is often necessary to avoid disputes.

Building codes in some major municipalities in the United States cater to the trade unions so much that the favorable economies of scale associated with manufactured, pre-assembled components are mitigated by increased labor costs. Installation costs escalate significantly when building codes require the rewiring and "hard" wiring of pre-assembled flexible lighting equipment to meet the "requirements" of the local code. There is considerable debate over whether such code "requirements" are predicated on valid electrical hazard issues or are "make-work" techniques guaranteeing additional work for the trade unions. Checking with the local codes prior to building systems' design and specification will enable the designer to accommodate unusual code requirements without significant labor costs to the building owner.

Flexibility

In order to maintain the usefulness of the building, the building system should provide sufficient flexibility to facilitate evolutionary changes in space utilization and layout. This is not to suggest that systems should be inordinately over-sized to anticipate future change. Rather it implies the need for space allowances (that is, space zoning) to facilitate and guide physical expansion of environmental system capacity and services in the future (fig. 8-13). It also implies the need for suitable accessibility into these service spaces.

When this approach is effectively developed in the initial design, the interior space arrangement can be physically organized and reorganized as necessary without the need for fundamental alterations in the service distribution network. Rather than providing all environmental services to each module of the interior, methods of bringing these services to any module should be designed.

The term *flexibility* includes several specific categories or modes of change that are discussed in the following sections.

Spatial Variety

This is the system capability that permits the designer to vary the permanent environmental characteristics of individual rooms. It is mandatory that room-to-room variations be possible without violating the basic structural, material, and dimensional discipline of the system.

Variations of this type most often respond to the specialized spatial demands of individual but related activities. Examples are distinctive system variations for corridors, reception rooms, private offices, and general offices, or for general classrooms, music rooms, and laboratories. There should be a variety of component and finish alternatives for selection, without deviating from the basic dimensional and design discipline that has been established.

As figures 8-16 and 8-17 illustrate, the continuous appearance of open-cell ceiling systems can be quite flexible in providing various types of lighting

Figure 8-16. Ceiling systems: modular open cell with accent lighting.

Figure 8-17. Ceiling systems: modular open cell with task-oriented fluorescent lighting.

equipment. These ceilings can also add variety and visual interest to the ceiling plane. It should be recognized, however, that for acoustical privacy, absorptive material should be used above the open cells, preferably affixed to the surface of the ceiling.

For a showroom type of environment, a selective pattern of downlights and adjustable luminaires provides appropriate highlighting (Fig. 8-16). Where a working environment is required, several ceiling cells can be removed and replaced with fluorescent luminaires (fig. 8-17). This task-oriented arrangement of lighting helps to focus attention to work areas, provides appropriate lighting to those work areas, and minimizes energy consumption.

Short-Term or Intermittent Changes

This category refers to situations that require conversion of space use with time and effort. This may involve temporary expansion or subdivision of a space, which implies some portability of space dividers and environmental devices.

Short-term variations may also involve changes in environmental conditions only, such as the changes in light characteristics that are required for an effective audiovisual space or the changes in thermal requirements associated with lecture rooms and other spaces that have variable occupancy conditions. In these cases, the system response should involve switches, dimmers, and thermostatic control elements that will permit automatic or human-activated response to changing environmental needs.

Open-plan office designs permit great flexibility, provided that the lighting, ventilation, power, communication, and electronics connections can accommodate the moving of workstations.

Long-Term or Permanent Changes

It is likely that some rearrangement of facilities and layout will take place during the life of the building. The system should facilitate such changes by permitting the necessary relocation of space dividers, ducts, supply and return air outlets, thermostatic and dimmer controls, acoustical and lighting devices, electric conduit and outlets, and control switches with a minimum of difficulty, disruption, and expense. This is not to infer that the system should be overdesigned but rather that it should be designed in a manner that will permit localized change with minimum labor, material waste, and disruption of adjacent areas (see fig. 8-13).

Expansion

Expansion procedures should be defined in the initial design. Where relevant, process equipment should be placed in a position that will facilitate and accommodate an orderly and logical phasing of expansion plans. Anticipated changes should require a minimum of demolition and disruptive interruptions within the existing building.

Furniture-Integrated Systems

The previous section dealt primarily with ceiling-building integrated environmental systems. As energy concerns increased through the 1970s and as the white-collar, service sector of our workforce grew, it became more of a challenge to provide proper environmental conditions with traditional building systems and furnishings. This challenge led to the development of modular furniture workstations that could be easily reconfigured to accommodate growth and change in the workplace. These workstations also provided an opportunity to introduce lighting in a specific task-oriented or task location manner. By localizing lighting equipment to areas when high light

Figure 8-18. Furniture systems: furniture-integrated lighting equipment provides both task and general lighting.

Figure 8-19. Architectural accenting or "fill lighting" (left wall).

levels were required for task performance, savings in lighting energy could be realized.

The advent of task lighting led to increased difficulties with brightness balancing, prompting the use of moderate-level, ambient (general) lighting. These two lighting subsystems are usually referred to as task-ambient. Figure 8-18 is an example of one such system. Here the ambient lighting is mounted on top of the furniture systems and indirectly lights the office area to a general level of 30 footcandles. The furniture integration of both lighting

Figure 8-20. Visually distant focus.

Figure 8-21. Atria spaces can serve as areas of visual relief.

subsystems enables most or all of the building electrical power to occur through the floor rather than splitting power to both floor and ceiling. This increased power requirement through the floor for lighting and the increased telecommunications and electronic task cabling needs of today's offices significantly impact the design of the floor system.

Furniture-integrated systems cannot always meet the lighting needs in a given building envelope. This is particularly encountered in renovation work where the building shell remains completely intact. Where ceiling heights are rather low (8 feet 6 inches or less), the most appropriate ambient light may be developed with a direct ambient lighting subsystem, as shown in figure 8-19. Ceiling-integrated parabolic luminaires provide low-brightness general illumination of 40 footcandles, with localized task lighting providing an additional 45 or more footcandles. Both direct and indirect ambient lighting systems properly utilized can provide satisfactory conditions for electronic office tasks.

For spatial definitions and visual interest, architectural fill lighting is necessary. The incandescent wall accents shown in figure 8-19 add sparkle and interest to an environment that might otherwise be somewhat bland due to the uniform, fluorescent ambient lighting required for proper illumination of electronic office tasks. Using accent walls or highlighted artwork (fig. 8-20) and exterior views can provide the visually distant focuses required for periodic eye muscle relaxation.

In some climates it may be preferable to develop interior atria areas to provide an area of visual relief year-round for interior-bound workers. As figure 8-21 shows, trees and sculpture can be devised to provide an area of relief. Lounge space surrounding such atria can further serve as informal breakout or interaction areas where employees can casually meet to exchange ideas and information. Recognize that the introduction of appropriate daylight and/or electric light is important in these areas to facilitate the maintenance of the plant growth and to provide a lighted environment visually different from work areas.

Index

Page numbers in *italics* refer to figures; page numbers in **boldface** refer to tables.